Multifunctional Bio-Based Lubricants

Synthesis, properties and applications

Online at: https://doi.org/10.1088/978-0-7503-3435-8

Multifunctional Bio-Based Lubricants

Synthesis, properties and applications

Edited by
Ashish K Kasar, Arpith Siddaiah and Pradeep L Menezes
Department of Mechanical Engineering, University of Nevada, Reno, NV, USA

IOP Publishing, Bristol, UK

ISBN 978-0-7503-3435-8 (ebook)
ISBN 978-0-7503-3433-4 (print)
ISBN 978-0-7503-3436-5 (myPrint)
ISBN 978-0-7503-3434-1 (mobi)

DOI 10.1088/978-0-7503-3435-8

Version: 20230301

IOP ebooks

British Library Cataloguing-in-Publication Data: A catalogue record for this book is available from the British Library.

Published by IOP Publishing, wholly owned by The Institute of Physics, London

IOP Publishing, No.2 The Distillery, Glassfields, Avon Street, Bristol, BS2 0GR, UK

US Office: IOP Publishing, Inc., 190 North Independence Mall West, Suite 601, Philadelphia, PA 19106, USA

Contents

Preface ix

Editor biographies xi

List of contributors xiii

1 Synthesis of bio-based lubricants **1-1**
Madhu Agarwal, Karishma Maheshwari and Neha Pal

1.1 Introduction 1-1
1.2 Evolution of lubricants and bio-based lubricants 1-4
1.3 Leverage associated with bio-lubricants 1-5
1.4 Properties of vegetable oil derived lubricants 1-6
1.4.1 Viscosity 1-6
1.4.2 Pour point, flash point, and fire point 1-7
1.4.3 Oxidation stability (OS) 1-8
1.5 Reaction mechanism 1-8
1.6 Tailoring the properties of bio-based lubricants 1-10
1.6.1 Esterification 1-10
1.6.2 Hydrogenation (HG) 1-11
1.6.3 Oligomerization 1-12
1.6.4 Epoxidation 1-12
1.6.5 Chemical and thermal modification 1-14
1.6.6 Structure tailoring 1-15
1.7 Obstructions associated with bio/vegetable-based lubricants 1-16
1.8 Resurgence and recent advances 1-17
1.9 Conclusion 1-17
References 1-18

2 Bio-based lubricants and their environmental sustainability **2-1**
Satish Vasu Kailas, Jeng Haur Horng and P S Suvin

2.1 Introduction 2-1
2.2 Closed loop concept of sustainability 2-4
2.3 Synthesis and evaluation of green cutting fluid 2-7
2.3.1 Toxicity evaluation of constituents 2-9
2.3.2 GC–MS analysis of commercial and GCF samples 2-13
2.3.3 Biodegradability and corrosion study of cutting fluids 2-16
2.3.4 Corrosion aspect of MWF (corrosion tests) 2-17

2.3.5 Bacterial growth study 2-20
2.3.6 Drilling and tool chip tribometer test 2-24
2.4 Conclusion 2-29
References 2-30

3 Tribology of multifunctional bio-based lubricants 3-1
Sougata Roy
3.1 Introduction 3-1
3.2 Sources of bio-based lubricants 3-2
3.3 Classification of bio-based lubricants 3-3
3.4 Types of tribological testing to evaluate the multifunctionality of lubricants 3-5
3.5 Tribological behavior: friction behavior, wear behavior 3-7
3.5.1 Tribological characteristics of pure vegetable oils 3-8
3.5.2 Tribological characteristics of additized vegetable oils 3-10
3.5.3 Tribological characteristics of bio-based ionic liquids 3-14
3.6 Summary and future research perspectives 3-24
References 3-26

4 Multiphase bio-lubricants for engine components 4-1
Dipen Kumar Rajak, D D Pagar and Pradeep L Menezes
4.1 History 4-1
4.2 Introduction 4-2
4.2.1 Friction, wear and lubrication for tribological surfaces 4-2
4.2.2 Friction 4-3
4.2.3 Wear 4-5
4.2.4 Lubrication 4-5
4.3 Tribological behavior of mechanical components in ICE 4-10
4.3.1 Gears 4-13
4.3.2 Cams and tappets 4-15
4.3.3 Piston rings 4-18
4.4 Challenges and future scope of bio-lubricants for ICE components 4-22
4.5 Conclusions 4-23
References 4-24

5 Machine learning to develop lubrication strategies **5-1**
Ashish K Kasar, Parker Hill, Soumya Sikdar and Pradeep L Menezes

5.1 Introduction 5-1
5.1.1 History of machine learning 5-1
5.1.2 Artificial intelligence versus machine learning 5-3
5.2 General ML tasks 5-3
5.3 Support vector machine 5-4
5.4 Genetic algorithm 5-5
5.5 Applications in tribology 5-5
5.5.1 Formulation and usage of lubricants 5-5
5.5.2 Wear condition monitoring 5-8
5.5.3 Value prediction 5-10
5.6 Conclusions 5-14
References 5-14

6 Environmentally friendly grease **6-1**
Ashish K Kasar, Reuel Baluyut, Jonfred Christian Diamzon, Morgan Lester, Rich Milani, Guillermo Ramirez and Pradeep L Menezes

6.1 Introduction 6-1
6.2 Types of grease and their application 6-2
6.3 Environmentally friendly grease 6-3
6.3.1 Biodegradability 6-4
6.3.2 Aquatic toxicity 6-5
6.3.3 Bioaccumulation 6-6
6.3.4 Labeling 6-7
6.4 Grease and regulation 6-9
6.4.1 Disposal of mineral and synthetic based lubricants 6-10
6.5 A comparison between environmentally friendly grease and conventional grease 6-11
6.5.1 Current environmentally friendly grease application and performance 6-12
6.6 Market analysis of PAG and synthetic esters 6-15
6.7 Conclusions 6-17
References 6-17

7 Additives for lubricants **7-1**
Ashish K Kasar, Logan Richardson, Alexander Scherp, Chris Tyree, Jacob Welty and Pradeep L Menezes

7.1 Introduction 7-1
7.2 Classification of additives based on functionality 7-2
7.2.1 Surface protective additives 7-2
7.2.2 Performance additives 7-7
7.2.3 Lubricant protective additives 7-11
7.2.4 Multifunctional additives 7-16
7.3 Conclusions 7-25
References 7-26

8 Environmentally friendly cutting fluids **8-1**
Ashish K Kasar, Alex Kandaras and Pradeep L Menezes

8.1 Introduction 8-1
8.2 Environmental sustainability 8-2
8.2.1 Composition of petroleum 8-4
8.2.2 Mineral oil toxicity 8-4
8.2.3 Composition of vegetable oils 8-5
8.2.4 Biodegradation comparison 8-6
8.3 Advantages of environmentally friendly cutting fluids 8-6
8.3.1 Rheological properties 8-8
8.4 Plant-based cutting fluids with extreme pressure additive inclusion 8-10
8.4.1 Plant-based nanofluids 8-11
8.5 Economic viability 8-13
8.6 Conclusions 8-14
Acknowledgements 8-15
References 8-16

Preface

Lubricating oils have been an integral part of engineering innovation in the fields of aerospace, automobile, and machining, etc. The main purpose of industrial lubricants is to reduce friction and wear of the component. However, the current innovation challenges require multifunctional lubrication where the lubricant not only provides desired friction and wear but also prevents oxidation, collects wear debris, and transfers heat for efficient energy transfer. The large usage of lubricants can be detrimental to the environment due to accidental or negligent disposal. To avoid this threat, environmentally friendly lubricants have been continuously developed. This book emphasizes various strategies for the development of environmentally friendly lubricants and their performance with respect to conventional lubricants.

Chapter 1 describes the three main categories of oils—mineral oil, natural oil, and synthetic oil while focusing on natural oils, mainly vegetable oils. The fundamental properties of vegetable oils are detailed to highlight the potential of vegetable oil-derived lubricants. Further, chemical modification processes are discussed to tailor the properties of vegetable oil-based lubricants.

Chapter 2 presents the framework of environmental sustainability with a closed-loop concept of sustainability. This chapter discusses various toxicity evaluation methods and the biodegradability of metalworking fluids. The frictional properties of green metal working fluids based on coconut oil and a variety of non-toxic emulsifiers are also discussed along with their environmental impact.

Chapter 3 focuses on the tribological properties of the various bio-based lubricants. First, the chapter details the various tribological testing methods to evaluate the multifunctionality of the lubricants followed by their frictional and wear properties. Two main categories of additives, namely, nano-particle additives and ionic liquid additives, and their effect on bio-based lubricants are discussed. This chapter also highlights the future research goals to effectively utilize nanoparticles as additives using machine learning and simulation-based software.

Chapter 4 addresses the application of multifunctional bio-based lubricants for automobile applications, particularly for internal combustion engines. This chapter also describes the functional concepts of friction, wear, and different lubrication regimes. The applicability of bio-based lubricants such as soybean oil, palm oil, sunflower oil, rapeseed oil, etc. for internal combustion engine components—gear, cams, tappets, and piston rings were discussed.

Chapter 5 describes the application of machine learning techniques to develop lubrication strategies. This chapter is aimed to highlight the utilization of computing power to deal with the complexity of tribological performance using various material properties, lubricant properties, and operating parameters to avoid catastrophic failures during operation. This chapter also discusses the use of artificial neural networks to develop a lubricant mixture using vegetable oils and additives.

Chapter 6 discusses the other important lubricant category, i.e., grease. Grease is a semi-fluid product of the dispersion of a thickening agent in liquid lubricant.

This chapter mainly focuses on environmentally friendly grease derived from vegetable oils, polyalkylene glycol, and synthetic esters. The biodegradation of greases made using these three sources is compared along with their tribological properties. Additionally, a market analysis of this environmentally friendly grease is also discussed in this chapter.

Chapter 7 details the wide variety of additives used for multifunctional bio-based lubricants. The lubricant additives are categorized based on their function which are surface protection, tribological performance, and lubricant protection through enhancing the physical properties of the lubricant. Various multifunctional additives are also discussed to enhance the applicability of bio-based lubricants.

Chapter 8 discusses the other important application of liquid lubricant which is cutting fluids. Cutting fluids not only reduce friction but also cool the machining region and take away the formed chips to enhance the life of the tool. This chapter suggests the necessity of developing environmentally friendly cutting fluid due to the formation of mist and micro-droplets that can cause severe hazards to the environment. The environmental cutting fluids are discussed in detail with respect to their chemistry, physical, chemical, and tribological properties along with their economic viability to develop sustainability.

This book is intended for professionals as well as university students to provide basic understating as well as the recent development in the field of multifunctional bio-based lubricants. This book has an extensive listing of results of different concerning fields concerning chemistry, mechanical, materials, and environmental sciences. This book also has extensive lists of references at the end of each chapter to provide an excellent source of references in the field of bio-based lubricants.

Comprehensive knowledge of multifunctional bio-based lubricants through this book has been possible with the collective efforts of various research groups around the world.

Editor biographies

Ashish K Kasar

Dr Ashish K Kasar received his PhD degree from the Mechanical Engineering Department at the University of Nevada, Reno. Prior to his PhD, He received his Master of Technology degree from Indian Institute of Technology, Kanpur in Materials Science and Engineering and Bachelor of technology from National Institute of Technology, Raipur. In his research career, he has published 36 articles in peer-reviewed journals, two book chapters. He has also presented his research work in prestigious conferences—Society of Tribologists and Lubrication Engineers (STLE), The Minerals, Metals & Materials Society (TMS) and The Materials Science & Technology (MS&T). His research interests include tribology, environmentally friendly lubricants, self-lubricating composites, corrosion, and tribo-corrosion.

Arpith Siddaiah

Dr Siddaiah received a PhD in Mechanical Engineering from the Mechanical Engineering Department at the University of Nevada, Reno. Prior to his PhD, he received his Master of Technology degree from Birla Institute of Technology, Ranchi, India. His research interests include tribology, tribocorrosion, corrosion, surface engineering, surface enhancements and texturing, laser-based tribological enhancements, nanocomposite coatings, and additively manufactured metallic materials. He has produced over 25 peer reviewed research articles , 14 conference publications, One book publication and one patent during his career. He has also presented his research work in prestigious conferences—Society of Tribologists and Lubrication Engineers (STLE), The Minerals, Metals & Materials Society (TMS) and The Materials Science & Technology (MS&T).

Pradeep L Menezes

Dr Pradeep L Menezes is an accomplished Associate Professor in the Mechanical Engineering Department at the University of Nevada, Reno. He has previously worked as an adjunct assistant professor in the Industrial & Manufacturing Engineering Department at the University of Wisconsin-Milwaukee and post-doctoral research associate in the Industrial Engineering Department at the University of Pittsburgh. He has a productive research career with over 160 peer-reviewed journal publications,

40 book chapters, eight books, and one patent. He currently serves as a reviewer on over 80 prestigious journals and an editorial board member of 10 journals. He has also participated in many national and international conferences in various roles. His research interests include tribology, bio-lubricants, corrosion, tribo-corrosion, composites, cold spray additive manufacturing, surface processing, stress corrosion cracking, nanocrystalline materials, and mechanical behavior of materials.

List of contributors

Madhu Agarwal
Department of Chemical Engineering, Malaviya National Institute of Technology, Jaipur, India

Reuel Baluyut
Department of Mechanical Engineering, University of Nevada, Reno, USA

Jonfred Christian Diamzon
Department of Mechanical Engineering, University of Nevada, Reno, USA

Parker Hill
Department of Mechanical Engineering, University of Nevada, Reno, USA

Jeng Haur Horng
Department of Power Mechanical Engineering, National Formosa University, Yunlin County, Taiwan

Satish V Kailas
Department of Mechanical Engineering, Indian Institute of Science, Bengaluru, India

Alex Kandaras
Department of Mechanical Engineering, University of Nevada, Reno, USA

Morgan Lester
Department of Mechanical Engineering, University of Nevada, Reno, USA

Karishma Maheshwari
Department of Chemical Engineering, Malaviya National Institute of Technology, Jaipur, India

Richard Milani
Department of Mechanical Engineering, University of Nevada, Reno, USA

D D Pagar
Department of Mechanical Engineering, K K Wagh Institute of Engineering Education and Research, Nashik 422003, India

Neha Pal
Department of Chemical Engineering, Malaviya National Institute of Technology Jaipur, Jaipur 302017, India

Dipen Kumar Rajak
Department of Mechanical Engineering, SITRC, Maharashtra, India

Guillermo Ramirez
Department of Mechanical Engineering, University of Nevada, Reno, USA

Logan Richardson
Department of Mechanical Engineering, University of Nevada, Reno, USA

Sougata Roy
Department of Mechanical Engineering, University of North Dakota, Grand Forks, ND, USA

Alexander Scherp
Department of Mechanical Engineering, University of Nevada, Reno, USA

Soumya Sikdar
Department of Mechanical Engineering, University of Nevada, Reno, USA

P S Suvin
Department of Mechanical Engineering, Indian Institute of Science, Bengaluru, India

Chris Tyree
Department of Mechanical Engineering, University of Nevada, Reno, USA

Jacob Welty
Department of Mechanical Engineering, University of Nevada, Reno, USA

IOP Publishing

Multifunctional Bio-Based Lubricants
Synthesis, properties and applications
Ashish K Kasar, Arpith Siddaiah and Pradeep L Menezes

Chapter 1

Synthesis of bio-based lubricants

Madhu Agarwal, Karishma Maheshwari and Neha Pal

Lubricating oils are extensively utilized in automobiles, machinery, aviation, refrigerators, agro-equipment, marine applications, etc and are well known to reduce the friction between two surfaces in connection. Lubricants primarily have base stocks with 80% oily composition with the traits comprising viscosity, pour point, and stability in supplementation of additives to tailor the properties. Because petroleum-derived lubricants are environmentally undesirable and depleting, exploration in the field of synthesizing bio-based lubricants has come into existence and gained attention. In the same pathway, the development of lubricating oils from completely eco-friendly agro- derived residues has recently been studied, eradicating the socioeconomic impracticality. Here, we first strategize the evolution of bio-based lubricating oils with their advantages over convention lubricants. Moreover, this chapter reviews the potential vegetable oils studies performed to operate within bio-lubricant specifications proposing the reaction mechanism of autoxidation. However, obstructions to the authors' opinion are stated, suggesting probable resistance concerning practical application.

1.1 Introduction

Due to rapidly increasing industrialization, petroleum resources are decreasing and, therefore, increasing economic costs [1]. Moreover, conventional resources like mineral lubricants are unviable because of toxicity and low biodegradability. This effect has paved the way for exploring bio-based lubricants that have pollution-free properties. Lubricants are antifriction chemical coatings applied to gaps in conjugation with two separated surfaces. A lubricant reduces wear and tear, minimizing machine failure and maintaining the process operation [2]. Significant properties of bio-lubricants responsible for preventing wear and tear between the two conjugative surfaces include viscosity, temperature, stability, corrosiveness, flammability, compatibility, and environmental friendliness. Moreover, toxicity and biodegradability also have significant implications as far as bio-lubricants are considered [3].

doi:10.1088/978-0-7503-3435-8ch1

Initially, some vegetable oils were reported for utilization as bio-lubricants. Some of the applications of vegetable oils for lubricants in the various field have been highlighted in figure 1.1 [2]. But recently, a preliminary investigation was carried out to enhance bio-lubricants' quality, notably via the presence of the catalyst and chemical modification. For instance, Tulashie and Kotoka (2020) analyzed the formulation of bio-based materials by chemical modifications of palm kernel, castor, coconut oils, and additives via transesterification, reporting the highest kinematic viscosity of castor oil lubricant to be 208 cSt and 16 cSt at 40 °C and 100 °C, respectively [4]. Other investigators, Almeida *et al* (2020), performed experiments using phaolefin oil yubase 8 and the modifications were attempted using solvent-free conditions and accelerated time of reaction, i.e. 20 min. This resulted in the highest viscosity of 7000 cP of cdn-OA synthesized with cardanol, acetic acid, and formic acid [5]. Also, another investigation performed by Soufi *et al* (2019) reported the re-utilization of waste cooking oil to develop a bio-lubricant, revealing an optimal 83% reaction yield and other related parameters, including a 1% catalyst loading and a reaction period of 30 min [6]. Therefore, it can be summarized that the properties of the bio-lubricants could be improvised with the inclusion of catalyst loadings, non-toxic chemical usage, and bio-derived substances [1].

Bio-lubricants are mainly synthesized using bio-derived oils. Various bio-oils and their applications are shown in figure 1.1. Some commonly reported literature includes the consumption of bio-derived oil, for instance, safflower oil, cuphea oil, tallow oil, rapeseed oil, canola oil, coconut oil, soya bean oil, olive oil, Crambe oil, sunflower, jojoba oil, linseed oil, etc. The most generalized prior chemical treatment for conversion of vegetable oils is the esterification of free fatty acids (FAs) with solvents like methanol in the existence of acid catalyst and consecutively processing of the oil for transesterification of the produced methyl esters to form bio-lubricants. However, several ways are reported for tailoring bio-lubricants properties: hydrogenation,

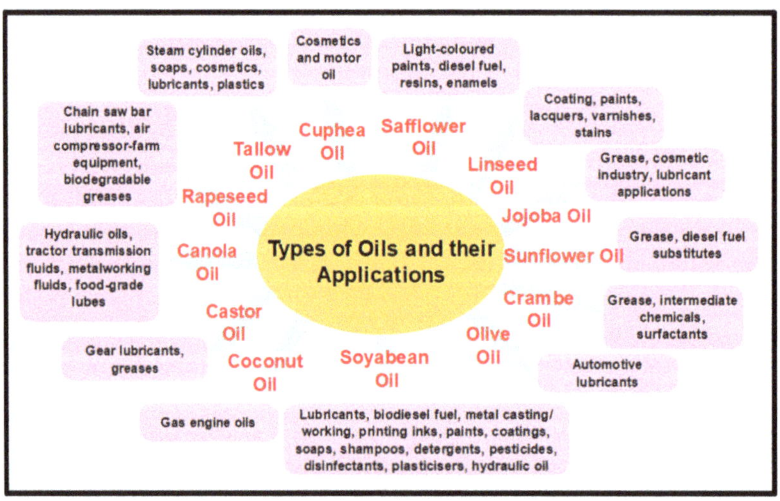

Figure 1.1. Types of bio-derived oils and their applications as lubricants in various fields.

oligomerization, epoxidation, chemical tuning, structure amendments, and thermal implementation approach, other than the esterification technique [7].

Jeevan and Jayaram (2018) examined the modification of vegetable oils from neem, mahua, and mineral oil via the epoxidation process revealing the improved lubricating effects with kinematic viscosity (KV) 21, 13, and 33 cSt at 40 °C, which led to the enhanced affinity between metal surfaces [8]. Another investigation by Mehdi *et al* (2019) synthesized a sustainable lubricant from pistacia Atlantica mutica oil using methyl ester to trimethylolpropane in the ratio of 5:1 under a vacuum pressure of 424.5 mmHg and a catalyst weight percentage of 1.45% resulting in the reaction yield of 82.5% [9]. Taha-Tijerina *et al* (2019) investigated the reinforcement of SiO_2 with sunflower, soybean, corn, and paraffinic oils, revealing the highest thermal conductivity enhancement of around 11% with nanoparticle incorporation of 0.25 wt. % in sunflower and soya bean oil, leading to 60% and 45% improvements in loading capacities [10]. A study analyzed Indian mustard seed oil with 1.1 wt.% KOH that leads to a high conversion percentage of 93%, ensuring the synthesized bio-lubricant is of industrial specifications [11]. In one of the studies, rapeseed and castor oil were examined as the base oil for bio-lubricant synthesis incorporating titanium isopropoxide, obtaining 93% conversion [12]. In another study by Wahyuningsih *et al* (2020), two novel chemicals (Dioxo-Dioxane and Dioxo-Dioxepane Ethyl Oleate) were synthesized as derivatives of bio-lubricants using *p*-toluene sulphonic acid as catalyst and reporting percentage yield of 93.9% via esterification [13]. Henceforth, it can be seen that chemical alteration and nanoparticles, additives, and other material composites enhance the performances of bio-lubricants withstanding a wide range of process temperatures by improving properties, such as solvency, biodegradability, low volatility, stability, viscometric properties, etc. From the evidence, it can be seen that base stocks with other incorporations enhance the performance of lubricating oils.

Additionally, recent investigations include a newer synthesis route of bio-lubricants incorporating fruit waste like banana peel with slippery touch and smooth textures that impart the lowest friction on the surface. Examinations were performed in which banana peels were blended on both sides, inner and outer, with paraffin oil [2]. Moreover, there are bio-lubricants synthesized from castor oil as it is non-edible [12]. Apart from this, there are petroleum-based oils and vegetable oils wherein the researchers concluded that such base stocks have improved friction and wear performance for high temperature application [13].

With the stated properties, base stocks, and modification, the present article aims at drawing attention to the evolution of lubricants based on sustainability while delineating the leverages associated with bio-lubricants. Moreover, the characteristic properties like viscosity, pour point, flash point, stability, fire point, etc, are discussed in detail, drawing out relativity with the bio-lubricants and recent advances. Also, the exact reaction mechanism is presented by proliferating the tailoring of bio-lubricants for better performance. Additionally, the difficulties in synthesizing, materials selection, applications, etc, are stated in the chapter. The recent advances presented in this chapter provide a systematic approach to developing bio-lubricants.

1.2 Evolution of lubricants and bio-based lubricants

Presently, several varieties of lubricants are being utilized at commercial scale and lab scale, such as grease, oil, and other functional fluid formulations with about 90% of the applications. Primarily, lubricants can reduce friction, transmit energy, protect against corrosion, disperse wear, block the entrance of foreign bodies, sometimes work as a sealant, etc. Due to high demands of lubricants, a trend has been developed for using natural, animal, and plant-derived materials for synthesizing bio-lubricants, for instance, palm, tallow, lard, etc which can reduce friction [13].

In 1859, the petroleum sector industry began the first commercial-scale oil drilling in the Titusville area in the USA. This examination over time developed a new interest in the field of natural oils to be consumed as natural oils. Natural oils have had greater resurgence than petroleum oils due to the point of detrition in the research area and the need for sustainability [14]. The wide variety of natural oils, including non-edible oils, makes the capital worth raising. The introduction of new chemical synthesis involving additive incorporation adds to significantly improved loading capability, lubricants properties, corrosion-resistance, and thermal stability. With a resurgence in this field for 150 years, bio-lubricants surpassed other lubricants due to the advantages of effort, time, and economics [10].

There are broadly three bifurcations based on feedstock: natural oils, synthetic oils and mineral oils (figure 1.2). Mineral oils are generally based on petroleum feedstock, and have been utilized for decades, whereas naturally derived oils include the base of the plant, animal, waste materials such as animal fats, tallow, bark, leaf,

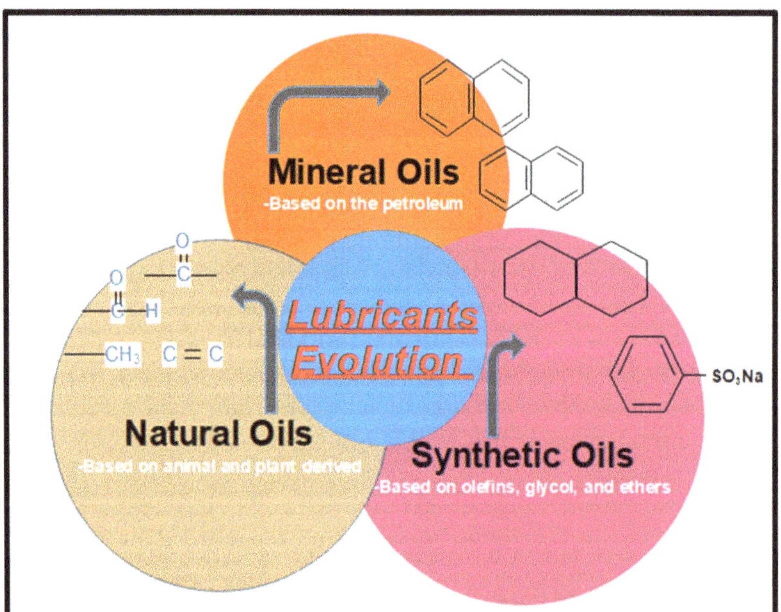

Figure 1.2. Lubricants evolution based on chemical base stock.

etc, and synthetic oils are chemical-based oils made up of olefins (polyalphaolefins), synthetic esters, polyglycols, alkylated aromatic compounds, ether (perfluoroalkyl polyether), etc representing a shift of the lubrication sector. It has been found to inculcate sustainability and conservation as ascribed statistics that bio-based lubricants have benign chemicals of hydrocarbon structure to be used as functional liquids in base stocks. Bio-lubricants render biodegradable chemical constituents with non-toxic characteristics to living beings and aquacultures. Therefore, the development of bio-derived lubricants has gained tremendous attention [12].

Due to significant issues in other lubricants, the target has been focused on the bio-lubricants as bio-derived fuels deterring the use of petroleum oils. In the global market, their evolution is gaining attention due to their environmentally friendly nature and cost-effective manufacturing without toxic effects and keeping in mind the stringent norms. Progressively, the need for bio-lubricants is due to increasing demand for sustainable materials, renewability, and feasibly socioeconomic alternatives [10].

Additionally, bio-lubricants have superior performance in lubricity and resistance to wear over other lubricants, facilitating the resurgence of usage in industrial applications. Lubricity with high values reduces frictional losses and gives enhanced efficacy, resulting in better economy and increased power outcomes. In contrast, they have relatively low thermal stability and higher pour point, leading to expansion of synthetic bio-lubricants [8]. Ultimately, these are bio-based feedstocks offering the advantages of high lubricity, leading to better fuel economy.

Additionally, low volatility results in decreased exhaust discharges, more remarkable shear ability, and good detergency, eliminating the need for additives. There are high dispersal properties, and the most rapid biodegradation is environmentally advantageous [3]. Other leverages are briefed in the upcoming section.

1.3 Leverage associated with bio-lubricants

Bio-based lubricants are derived from vegetable oils already ascribed in the above part of the chapter. These are extracted from the seed or organic fruits like soya bean, rapeseed, olive, peanut, palm oil, castor oil, etc, which are commonly investigated as feedstock for developing bio-lubricants. These, in comparison, have the following advantages over other lubricants:

(i) Ability to have renewable characteristics: toxicity associated with bio-lubricants is very low, or it can be said that it is negligible to aquaculture, with potentially good lubricating properties.
(ii) Biodegrading ability: the biodegradability of the bio-based substances is primarily high in the range of 90%–98%, whereas mineral oils have huge non-biodegradable properties.
(iii) Higher viscosity index (VI): bio-lubricants possess high VI because of the solid intermolecular forces between double bonds, leading to high affinity and interactions. Also, the molecule has a linear chemical structure because VI is high.

(iv) Low rate of evaporation: bio-lubricants have approximately 20% lower evaporation rates than mineral-based oils and synthetic oils.

1.4 Properties of vegetable oil derived lubricants

It is essential to understand the physicochemical properties of lubricants derived from bio-based materials. The properties of vegetable-based oils are connected to the structural aspects of fluid particles. Vegetable oils usually have many advantages, such as high biodegradability, less toxicity, high flash point, high viscosity, poor friction, and wear relative to mineral oils [1, 15]. The performance of vegetable oil as a bio-lubricant depends on various properties such as viscosity, stability, flash point, and pour point.

1.4.1 Viscosity

The viscosity of oil measures the flow resistance of the material. High viscosity implies that the material has high flow tolerance, while less viscosity shows less flow tolerance [16]. The oil's viscosity measures its resistance to shear stress or tensile stress to progressive deformation. They are the functional features of every lubricant. Viscosity defines the hydrodynamic conduct of lubricants and calculates oil tolerance to shear. It is measured as shear tension, shear speed, or time taken to move through a typical hole at a specific temperature by an oil sample. Viscosity specifically determined by shear is measured in centipoise (CST) as complex viscosity. However, kinematic viscosity is the most popular term used in the oil industry, which is measured as complex viscosity separated into oil density of 313 K and 373 K. It is measured in centistokes ($1\ cSt = 1\ mm^2\ s^{-1}$). VI is a non-dimensional number that shows the rate of change in the viscosity of a lubricant with a temperature change. Therefore, it defines the temperature range of a working lubricant. Bio-lubricants must have a high VI to maintain consistent viscosity with temperature rise. Viscosity plays a crucial function in eliminating discomfort and wear by the lubricant. High viscosity improves the oil temperature and drag, while relatively low viscosity improves the interaction with the moving parts of fuel and concrete. One of the considerations influencing a lubricant's viscosity is the length of the carbon chain. As the duration of the FA chain rises, the level of spontaneous intermolecular interactions rises, which increases viscosity [17]. However, the implementation of branching results in a more compact molecular configuration of chemically modified vegetable oils [7]. Another aspect that influences the lubricant's viscosity is the degree of unsaturation [18]. Viscosity is correlated explicitly to temperature as well. Bio-lubricants with high VI have less viscosity when the temperature is higher. Bio-lubricant branching often influences the VI when increased alcohol branching or carboxylic acid facilitates a lower VI. In contrast, increasing duration has a detrimental impact by increased VI.

A significant number of lubricants based on mineral or synthetic oil are identified in the market that covers a broad range of viscosities, prioritizing their use for a wide variety of applications. For example, at 100 °C, kinematic viscosity of 3.8–20 cSt, 4.1–40 cSt, or 35.2–3520 cSt are required for engine oils, lubricants for automotive gear, and manufacturing gear, respectively [18]. Quinchia *et al* (2009) produced a bio-lubricant having superior thermal sensitivity and kinematic viscosity. Sunflower oil containing high-oleic acid was blended with polymer additives. Styrene–butadiene–styrene and ethylene-vinyl-acetate polymer additives are used at various percentages (0.5%–5% w/w). The maximum kinematic viscosity at 40 °C and 100 °C was enhanced at approximately 150–250 cSt and 26–36 cSt, respectively. Nevertheless, the concentration of copolymer in sunflower oil was minimal because a certain degree of sedimentation, cloudiness and additional developments were therefore not practicable [19]. Therefore, for improving bio-lubricant performance, Quinchia *et al* (2010) used ethylene-vinyl-acetate copolymer as a viscosity modifier for different kinds of vegetable oils (sunflower oil (SO), high-oleic sunflower oil, castor oil, and soybean oil) theoretically applicable as lubricants. Results show that the most relevant viscosity enhancement was observed at moderate temperatures for low-viscosity vegetable oils such as soybean oil, sunflower oil, and high-oleic sunflower oil. Viscosity increased from 330% to 420% at 40 °C as compared with the original oil obtained. The lowest increments led to castor oil/ethylene-vinyl-acetate blends, mostly at low temperatures [20]. Tulashie and Kotoka *et al* (2020) tested chemical alteration and bio-lubricants development by transesterifying coconut oil, castor, palm kernel oil, and special additives (SAE 40). The physical–chemical characteristics of bio-based lubricants were analyzed and related to SAE 40. At a temperature of 40 °C and 100 °C, modified coconut-based bio-lubricant kinematic viscosity is 42.43 cSt, and 10.11 cSt, respectively. Castor-based bio-lubricant is 208.39 cSt, and 16.47 cSt; palm kernel-based bio-lubricant is 58.49 cSt, and 11.00 cSt; and SAE 40 is 170.45 cSt and 15.60 cSt, respectively [4]. The features of bio-lubricants characteristics show that vegetable oils are strong comparative candidates for bio-lubricants for motor engines if properly modified [4].

1.4.2 Pour point, flash point, and fire point

The performance of a bio-lubricant is evaluated by its characteristics, such as its pour point (PP), flash point (FP), and fire point (FiP). PP is the lowest temperature at which oils (lubricants) cease to flow. To maintain its fluidity at low temperatures a low PP of any lubricant is desirable. The FP of a lubricant is the lowest temperature at which oil creates vapor that flashes when exposed to flame. On the other side, FiP corresponds to the temperature at which oil burns continuously upon exposure to flame. The volatility, fire tolerance, and storage temperature specifications for lubricants are calculated by the FP and FiP [3]. Low-temperature bio-lubricants need low PP, FP, and FiP. PP is the lubricant's significant low-temperature characteristic as the lubricant becomes semi-solid at the lowest temperature and lacks flux products. Overall, low temperature lubricants are ideal since these lubricants are suitable for vacuum lubrication at exceptionally minimal temperatures [21] and in cold weather conditions [16]. If the lubricant does not flow easily, machinery malfunction is considerably higher. In the past years, various investigations have been done to establish the lower temperature characteristics of vegetable oil. These studies found that with long-term exposure to a cold environment, most vegetable oils turn cloudy, precipitated, and firm at −10 °C, resulting in low flow and pumping [22]. It is due to the macro-crystal structures of the vegetable oils at low temperatures. These frameworks impair the system's simple flow by loss of kinetic energy during the self-stacking of the molecules [8].

Vegetable oils are restricted to directly used lubricants in vehicles due to low stability and poor pour point. Hence, various researchers are working to improve the performance of lubricants at low temperatures (PP) performance. Zhang *et al* (2020) altered oleic acid through epoxidation and further esterification. Epoxidation is a primary method for changing unsaturated FA to manufacture lubricants. In this study, improved bio-lubricants provided excellent VI (>116) and low-temperature efficiency (pour point <−48 °C). Therefore, proper chemical modification is required to improve the low-temperature performance of plant-based lubricants [23].

1.4.3 Oxidation stability (OS)

Oxidation stability (OS) is a chemical reaction that occurs when the lubricant comes into contact with oxygen; it undergoes an oxidation chemical reaction. The capability of a lubricant to resist oxidation is referred to as OS. Numerous aspects influence the rate of lubricant oxidation, including the presence of water, pressure, and temperature [16]. Additionally, as the temperature rises, the lubricant's lifespan declines. Strong OS is a crucial criterion for lubricants because poor OS causes the lubricant to oxidize quickly if untreated. As a result, the lubricant polymerizes and has plastic-like consistency [24]. Due to unsaturated FA, the OS of vegetable oil is considerably lower than that of synthetic esters [3]. The polymerization because of oxidation reaction enhances the viscosity and reduces lubricant functionality [1]. Deterioration causes the lubricant to polymerize which is unfavourable because it becomes volatile and corrosive. Also, its structural characteristics decline. In vegetable oil, OS is evaluated by primary FA [12]. Moreover, alkenyl chains are double-bound and react with an oxygen atom to develop free radicals, which degrade into peroxides and FA, leading to polymerization and fragmentation [25]. According to Zainal *et al* (2018), oxidation is affected by the degree of unsaturation of FA [26]. The oleic acids, linoleic acid, and linolenic acid are the primary FA with a double bond and an oxygen absorption rate of 1:100:800 [26–28]. Consequently, a high level of unsaturation causes oxidation to occur at a higher rate [25]. As per Gryglewicz *et al* (2003), a large amount of polyunsaturated FA found in rapeseed oil becomes less resistant to oxygen and high temperatures. It has been found that the kinematic viscosity of Jatropha oil escalates significantly because of its poor OS. The β-CH group is smoothly eliminated from its molecular structure, weakening the intermediate C–H bond. As a result, carboxylic acid is formed, degrading the lubricant [29].

1.5 Reaction mechanism

There are two proliferations for bio-lubricants: triesters and monoesters. The family of triesters comprises widely available oils and they are also named triglycerides, comprising three continuous FA cycles of chains attached with other functional groups like hydroxyl species. The oils are generally attached to these groups with ester bonds. Monoesters are generally smaller sections with similar chains, but other alcoholic chains impart different chemistry amongst molecules.

FA and alcohols have 14–22 large chains of hydrocarbons with varying degrees of saturation and unsaturation. Vegetable oils consist of different ranges of polar groups and non-polar sides, which leads to amphiphilic characteristics. These characteristics allow investigators to use them as hydrodynamic lubricants. Moreover, due to the long chemical groups of oils, the structure imparts low volatility in them, and the molecular weight of the oils leads to potentially strong characteristics.

Structural distribution, including triglycerides, is mainly accountable for irregularities in vegetable oils as bio-lubricants. Investigations state the initial point of unsaturation bonds that act as actives sites for new linkages to draw out the mechanism amongst the oils. Besides this, the associated thermal stability of oxidation states and imperfect properties like low temperatures have also been observed. These occur due to the double bond linkages in the structures of hydrocarbons. Moreover, investigators have drawn out the reactions based on the interactions between active sites. For instance, Soni and Agarwal (2014) have elaborated on the mechanism of vegetable oils, and the reaction representation of the oil autoxidation mechanism is shown below [1].

$$\text{Initiation: } R'H \rightarrow R'^{\cdot} + H^{\cdot} \tag{1.1}$$

$$\text{Propagation: } \begin{array}{c} R'^{\cdot} + O_2 \rightarrow R'OO^{\cdot} \\ R'OO^{\cdot} + R'H \rightarrow R'OOH + R'^{\cdot} \end{array} \tag{1.2}$$

$$\text{Branching: } \begin{array}{c} R'OOH \rightarrow R'O^{\cdot} + O^{\cdot}H \\ R'O^{\cdot} + R'H + O_2 \rightarrow R'OH + R'OO^{\cdot} \\ O^{\cdot}H + R'H + O_2 \rightarrow H_2O + R'OO^{\cdot} \end{array} \tag{1.3}$$

$$\text{Termination: } \begin{array}{c} R'OO^{\cdot} + R'OO^{\cdot} \rightarrow R'OOH + O_2 \\ R'OO^{\cdot} + R'^{\cdot} \rightarrow R'OOH \\ R'^{\cdot} + R'^{\cdot} \rightarrow R' - R' \end{array} \tag{1.4}$$

$$\text{Peroxide decomposition: } R'OOH \rightarrow \text{lower molecular chemical species} \tag{1.5}$$

$$\text{Polymerization: } R'OOH \rightarrow \text{lower molecular chemical species} \tag{1.6}$$

Vegetable oils especially have oxidation initiated by forming free radicals. These radicals are easily formed by removing the H atom from the methylene structure around the double bond. Free radicals readily react with oxygen resulting in peroxy radicals. Peroxy radicals then focus on another molecule of lipid to emit hydrogen atoms, resulting in the propagation of the oxidation process. After that, oxidative reaction and degradation tend to increase viscosity. Moreover, the alkene groups present in vegetable oil transform a molecule into other stable, functional groups, resulting in improved oxidative and thermal performances.

1.6 Tailoring the properties of bio-based lubricants

1.6.1 Esterification

Esterification is a reaction in which the three moles of methanol with acid or base catalyst leading to methyl ester and glycerol, as mentioned below, are made to react with triglyceride [30, 31]. Various types of alcohol used in the transesterification method are methanol, ethanol, propanol, butanol, and amyl alcohol. The two widely included in this chemical reaction are methanol and ethanol. The chemical reaction of the transesterification process is shown in figure 1.3. The primary explanation is its superior benefits of strong oil solubility, rapid triglyceride reaction rates, decent physical and chemical properties, and low cost.

Esterification will overcome the existence of H atoms in β compared to the glycerol OH group. Most typically used bio-lubricants are polyol esters, based on poly-alcohols such as neopentyl glycol (NPG), acid linear fatty medium/long-chain (C8–C18), trimethylolpropane (TMP), and pentaerythritol (PE) [31]. Polyol esters shape various bio-lubricant foundation fluids with multiple features based on their chemical composition. Zainal *et al* (2018) reported that the number of OH groups influences the esters' viscosity in the following manner: NPG < TMP < PE. The number of carboxylate groups within the ester molecule increases thicker grain film in high-load, thermal properties and may retain high-slip contacts. The viscosity of TMP esters is also greater than that of NPG esters, owing to the inclusion of three acid groupings in the TMP ester structure and the nature of only two groups in NPG ester molecules [26]. Sripada *et al* (2013) produced synthetic biological lubricants using Canola biodiesel and methyl oleate transesterification. Sodium methoxide has been utilized as a catalyst. A bio-lubricant of Canola biodiesel and methyl oleate has a strong percentage point with a value of −66 °C and −51 °C [33]. The polyunsaturation and unsaturation FA of bio-lubricants are produced from methyl canola biodiesel and oleate biodiesel. The reaction usually includes 3 mol of alcohol and 1 mol of triglyceride. However, a surplus amount of alcohol boosts the alkyl ester yield and facilitates its phase dissociation from the shaped glycerol [34]. The transesterification is affected by various parameters, including the nature of the catalyst (alkaline or acid), alcohol-to-oil molar ratio, and reaction temperatures.

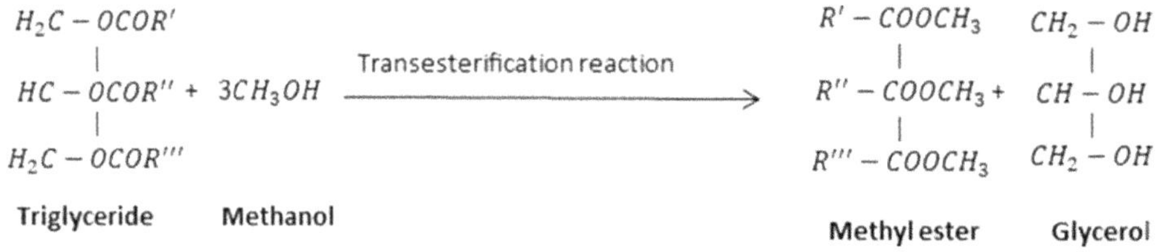

Figure 1.3. Transesterification process. Reprinted from [32], copyright (2013), with permission from Elsevier.

Catalyst selection is an increasingly significant element in the processing and producing of high-quality bio-lubricants.

The acid-based catalyst process is not commonly practiced in industrial applications relative to the base-catalyst transesterification process. One of the key reasons is that the acid-catalyzed transesterification process is around 4000 times longer than the base-catalyzed transesterification process. Acids, preferably sulfonic, phosphoric acids, sulphuric, and hydrochloric, which may contribute to corrosion of reactors, are catalyzed in this transesterification process. Although the product yield is high, the reaction rate is sluggish and it typically takes temperatures above 100 °C and more than 3 h to achieve complete conversion. Excess alcohol leads to alkyl esters formation. An unnecessary volume of alcohol renders it impossible to extract glycerol; therefore, it is necessary to allow the best possible use of alcohol in the oil [35].

The suggested base-catalyzed transesterification process notes that esters form an anionic intermediate in the presence of bases and may dissociate themselves from the original ester or form a new ester. The base catalysts corrode industrial machinery, which restricts its usage. The commonly used alkaline catalysts are sodium hydroxides, potassium hydroxides, carbonates, alkaline metal alcohols, and alkoxides (methoxide, ethoxide, isopropoxide, and butoxide) [31]. Among these alkaline metals, alcohols are the most potent catalysts since they produce relatively large amounts (> 98%), even though at low molar concentrations (0.5 mol%), in a short period of reaction. They need a shortage of water which does not render them ideal for traditional industrial processes [16]. Alkaline metal hydroxides (KOH and NaOH) are cheaper but less potent than metal alcohols. They can only achieve high conversions of esters by growing their concentration from 1% to 2%.

1.6.2 Hydrogenation (HG)

Hydrogenation is a reaction mechanism wherein the hydrogen in the triglycerides of the oil molecule is incorporated into the C=C bonds [36, 37]. In the HG process, three chemical reactions occur simultaneously: (1) geometric (isomerization); (2) location isomerization; and (3) double bond saturation [7]. Hydrogenated oil's physical and consistency feature is by the quantity of dual bonding and the FA isomers. In commercialized hydrogenation, nickel (Ni) catalysts mounted on Raney Ni, alumina, and silica, aluminium alloy are commonly utilized [36, 38]. However, the concentration of Ni trace in the crude contributes to needing to investigate a novel catalyst, due to concern about the toxicity. Therefore, the usage of novel metals such as Pd for the catalyst are desirable because Pd catalytic output is higher than Ni [36]. A catalyst's selectivity applies to the catalyst's capacity to minimize linolenic acid favourably before oleic acid and linoleic acid. Selectivity also applies to a catalyst's ability to decrease monomers by HG without decreasing to maximum saturation. This is regarded as partial HG, an essential way of avoiding oxidation of the lubricant's low-temperature properties such as pour point [36]. The selective HG process of ethyl esters in low-temperature sunflower oil was tested with assisted Pd, Pt, and Ru catalysts. The strongest catalytic efficiency of the Pd catalyst has been demonstrated with the selectivity of ester (C18:1) enhanced by the alteration of copper and lead catalysts or by introducing amines to the reaction medium. Moreover, the isomer's oleic acid (C18:1) benefits from being stable in an O_2 atmosphere that prevents polymerization [34].

1.6.3 Oligomerization

Oligomerization is simply an alteration of the two bonds of unsaturated FA, which are connected to two or more FA molecules. Estolides are produced where one fatty acid has a carboxylic acid that connects to another FA site for unsaturated oligomeric esters [39, 40]. The secondary ester relations to hydrolysis are more robust than triglyceride oils/fats and have more substantial physical properties for bio-lubricants. The degree of the molecule oligomerization or the average amount of FA based on FA is measured by the number of estolides [41]. Oxidation, polymerization, and oligomerization reaction are responsible for removing double bonds due to the variation in the viscosity of lubricant-based oils. Excessive chains can minimize biodegradability and increase acidity. The resulting solid from the internal esterification of unsaturated FA is an exciting category of compounds since the carboxyl acid category of one FA is added by the acid-catalyzed method to the unsaturated carbon of another FA. Poly-alpha-olefin (PAO) is now considered a high-performance lubricant developed by oligomerization of olefins that have been catalyzed by chromium, magnesium, titanium, molybdenum, and nickel oxides systems. The oligomerization of vegetable oil methyl linoleate is an ideal way to boost the bio-lubricant's oxidative stability and low fluidity [42]. Polybutene with very low viscosity may substitute a proportion of the hydrocarbon diluent. These low-viscosity polybutene oligomers have a neutral impact on smoke and lubricity but seem to have a beneficial effect on oil detergence and the blocking of low fatigue structures. Essential inventories may be processed utilizing several methods like distillation, solvent refining, hydrogen production, oligomerization, esterification, and reprocessing, although not restricted. Refined stock will be significantly free from manufacture, pollution, or direct materials usage.

1.6.4 Epoxidation

The word epoxy can be described as cyclic ethers consisting of three elements in the epoxy ring. Epoxidation is an essential reaction *in situ* for unsaturated FA. These double bonds in vegetable oils are used as a reactive site and may even be worked by epoxidation [15, 43]. This unsaturation is a downside for vegetable oil and restricts its usage as a lubricant when used at high temperatures. The usage of epoxy-identified vegetable oil in recent years has become more popular. Besides, plasticizers and polymer PVC additives extracted from vegetable oil have demonstrated better efficiency in terms of high heat and light tolerance. Epoxidized oil includes epoxy groups or rings of oxirane. The ultimate method of the epoxy group's synthesis is considered an epoxy reaction in which an alkene is reacted with organic peroxide acid.

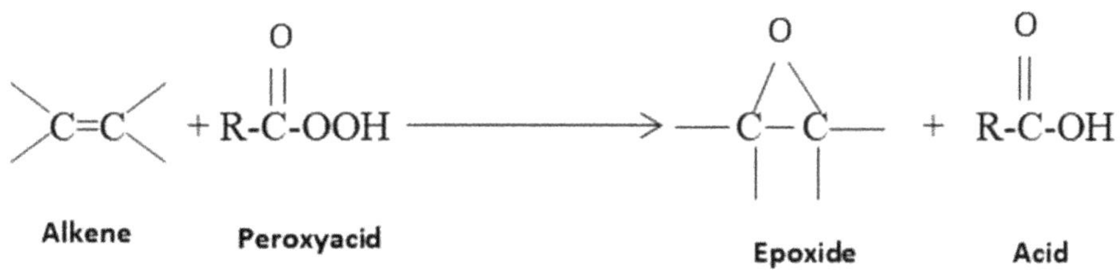

Figure 1.4. Reaction mechanism of peracids with double bonds.

As seen in figure 1.4, an alkene is usually reacted with peroxide acid to synthesize the epoxy groups or the oxirane circles. Wu *et al* (2000) used a traditional form of epoxidation (i.e., *in situ* generations of peracetic acid) of rapeseed oil (RP). In the presence of antioxidants, the oxidation stability of the epoxidized rapeseed oil improved dramatically. Epoxidized RP has more significant friction reduction and intense pressure than RP because the three-member ring formed by tribo-polymerization forms a polyester or polyether content film [44].

Epoxidation has gained significant attention because the C=C relation to the oxirane ring may be changed in a number of reactions under moderate conditions. Following the epoxidation reaction, the epoxidized oil would be ring-opened as well as the esterification process. Hwang and Erhan *et al* (2001) used sulfuric acid as a catalyst in the same string. The resulting OH group was esterified with an acid anhydride. The findings revealed that the oxidation stability has increased, and the pour point is significantly decreased by branching implementation [25]. In another study by Sharma *et al* (2013), the synthesis of a bio-based lubricant was accompanied by a one-step method involving the simultaneous opening of epoxy rings and the esterification of ethoxylated canola vegetable oil. This study uses a novel sulfated Ti-SBA-15 catalyst in the presence of acetic anhydride. The findings showed that tribological characteristics of ethoxylated canola oil are strong and obtained outstanding bio-based lubricant [33]. Chemical lubricants are created from epoxidized vegetable oils by cationic ion exchange resins [33]. The method includes ring-opening reactions, accompanied by esterifying the resulting hydroxyl group with different alcohols. A researcher, Arumugam *et al* (2012), reported the comparative study of epoxidized rapeseed oil and raw rapeseed oil. The findings obtained from the rotary bomb oxygen testing technique indicate that epoxidized rapeseed oxidation stability for natural rapeseed oil is twice the value. Moreover, the pour point of epoxidized rapeseed oil (−15 °C) is less than that of raw rapeseed oil (−11 °C).

Furthermore, the probable reactions during the epoxy ring-opening contribute significantly. Estolides are branched esters that are formed by linking the carboxylic acid group of one FA and oxidant ($HClO_4$) to a position of unsaturation of another FA (via carbocation). Nucleophile attacks by other FAs may result in an ester, namely an estolide. These ramified esters are hydrolytically more robust than triglycerides. The duration of the hydrocarbon chain and estolide number was also found to be important to the physicochemical characteristics of lubricants, as shown by the estolides obtained by 2-ethyl hexanol, oleic acid, and lauric acid [21].

1.6.5 Chemical and thermal modification

Vegetable oils have several attractive qualities but are not commonly employed as a base lubricant without modifications. Most vegetable oils have adverse physical properties, strong melting points, and poor thermal-oxidative stability. Vegetable oils have distinct structures and properties relative to mineral oils [45]. The presence of unsaturated FAs oxidizes vegetable oil more efficiently. Whereas mineral oils comprise mainly saturated linear, rammed, and cyclic hydrocarbons, vegetable oils are made up of carboxylic triacylglycerols in hydrocarbon chains with saturated bonds [45, 46]. These systemic variations appear in positive and negative ways. Natural triglycerides are biodegradable quite easily and are extremely efficient lubricants. However, lubrications derived from vegetable oil are partially reduced since they appear to have low oxidative stability. However, these drawbacks may be eliminated by chemically manipulating oils or synthetic esters partly obtained from sustainable sources [46].

Vegetable oil is used explicitly as base inventories, or lubricant additives are ineffective. It is less thermo-oxidative, has more viscosity, and is less volatile. It burns and deposits incompletely onto engine fuel injectors due to a strong viscosity and low volatility (about 11–17 times greater than diesel fuel). Therefore, it requires chemically processing before utilization as biofuels/bio-lubricants. This can be achieved in carboxylic reactions of vegetable oils, triglycerides/FAs/esters, or the FA chain olefin reactions. Transesterification of vegetable oil is the first phase in biofuel/lubricant processing. It generates FA alkyl esters along with glycerol and it can be used as biological fuel/bio-lubricant. It is transformed into its derivatives through one or more steps with various reactions to generate different forms of lubricants/additives depending upon industrial needs. Glycerol extracted from triglyceride esters and other essential chemicals can also be used as water-based lubricant additives.

The olefin functions (epoxy, carbohydrate, malleating, acrylation, hydroxylation, hydroformylation, and polymerization) perform in different reactions to create plasticizers polyurethanes, adhesives, paints, coatings [47]. The base stock for the preparation of environmentally safe, readily biodegradable lubricants may be used by unmodified vegetable oils and modified vegetable oil esters. There are structural esters of glycerin and long-chain (triglycerides) FAs. In all vegetable oils, the alcohol portion (glycerol) is similar. The FA in natural vegetable oils varies in chain length and double bond number. Natural triglycerides are biodegradable quite easily and are highly efficient lubricants. Nevertheless, their thermal, oxidation and hydrolytic stability are reduced. Pure vegetable oils are also only used in low thermal stress applications. The systemic 'double attachment' of the FA portion and the 'β-CH portion' of the alcoholic components is the source of thermal and oxidative instability of vegan oils. In alkenyl strings, double bonds react with oxygen in the atmosphere. The molecular framework will effectively remove the β-hydrogen atom, which allows the esters to cleave into acid and olefin. Another vulnerability of natural esters in the presence of water is their ability to hydrolyze [45, 47]. The thermal, oxidative, and hydrolytic stability of vegetable oils may mainly be enhanced by chemical modifications [45].

1.6.6 Structure tailoring

Structure tailoring has typically shown that the coefficient of boundary friction declines with the rising duration of the carbon chain. A long FA hydrocarbon chain is a significant molecular obstacle [48]. The period of the FA chains, which raises the covered region, leads to the rise in adsorbed film thickness. The temperature of FA in friction transfer rises with increased minimum FA chain length and molecular weight. However, short FA chains ($n = 8$) do not enhance lubrication; e.g., addition of short-chain FA into the ester synthesis showed decrement in polyol ester viscosity and flash point. The lubricity can also be enhanced by reducing the branching degree (to be more linear) of the base oil [49]. If the chain length rises, the durable wear rate of trimethylolpropane, and pentaerythritol linear chain acids declines, as shown in figure 1.5.

1.6.6.1 Conversion into cyclic carbonates of the olefinic functional groups and use of different nucleophiles to open the rings

Cyclic carbonates are an appealing green medium that is non-toxic, biodegradable and can be conveniently prepared with epoxy fatty esters in the presence of carbon dioxide.

The synthesis of FA derivatives/epoxidized oils into cyclic carbonates has two major advantages. Firstly, fresh products are prepared from natural resources;

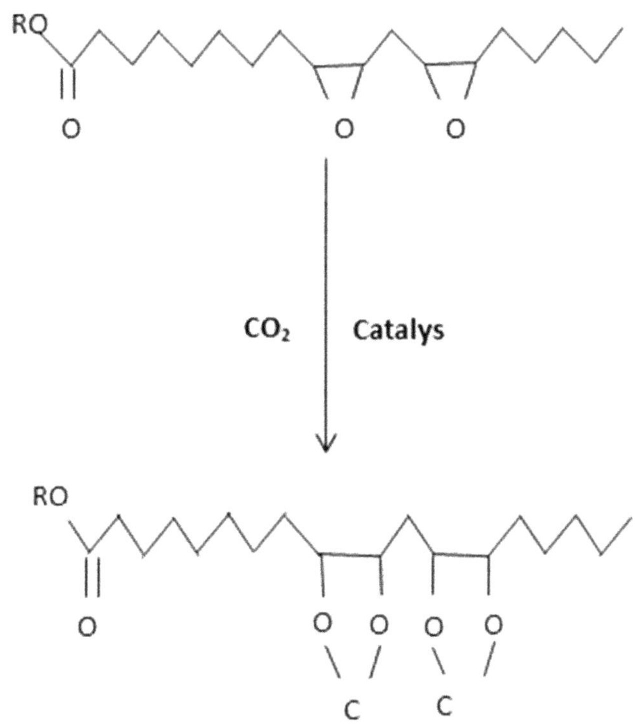

Figure 1.5. Carbonation of epoxy fatty esters.

secondly, carbon dioxide is recycled and included in the supply chain. The epoxy pair reaction with CO_2 is a research area which has been intensively studied and numerous homogeneous and heterogeneous catalysts have been established.

1.6.6.2 FA/ester epoxidation and its application as a bio-lubricant
The epoxidation of its olefin functionalities is one of the most essential measures to render a bio-lubricant (base and/or additive). The epoxidation of free FA/esters/ vegetable oils can happen by varying reaction parameters in the presence of peracids: reactant molar proportions, temperature, solvent character, presence or absence of a catalyst (mineral acids/ion replacement resins), stirring speed, and peroxy acid (peracetic, performic, m-chlorophenzoic acid) [53].

1.7 Obstructions associated with bio/vegetable-based lubricants

Although sustainable, vegetable-based oils have some limitations, and researchers concluded that bio-lubricants have the primary concern of poor resistance in terms of oxidative and thermal aspects. There is a presence of acyl group in the chemical structure due to which the oil is unstable, i.e. the bonding of glycol in oil leads to the formation of the tertiary β-hydrogen bond [7, 50]. Also, it has been reported that bio-lubricants have low-temperature characteristics because of the resulting macro structure limiting the flow behaviour due to kinetic energy loss via stacking [50]. However, in some cases, the ester functional group's presence is also responsible for the bio-lubricants being less volatile at high temperatures.

Recent reviews have delineated that bio-lubricants possess sustainability, biodegradability, pure nature, poor acidity, high VI, homogeneity, low parameters like pour point and iodine value, high oxidative stability and compatibility with elastomers [7]. Hence, being this advantageous, bio-lubricants possessing disadvantages like low temperature and poor resistance, etc have gained enormous attention due to the recent resurgence in this field. A new commercialisation horizon has opened the pathway toward more recent investigations [8]. Also, there are a few technical difficulties with bio-lubricants which need attention. From the author's point of interest, the problems can be overcome with the points noted below:

(i) A cost-effective quality test system needs to be developed.
(ii) There has been negligible focus given to the oxidation effects of oil on the system performance and the durability.
(iii) Discharge monitoring and testing with ample feed base stocks.
(iv) The investigations have limited focus on additive compatibility, material selection, stability, and long-term storage.
(v) One can focus on the continuous testing of the degree of oxidation, emissions, and several machine systems.
(vi) Environmental advantageous possessed by vegetable oil over other lubricating oils are needed to be marked for the purpose of commercialization.

(vii) Global markets need to keep in mind with cost balances and availability of the oils.
(viii) Development of newer additives with improved properties of stability and compatibility are the major concern.

1.8 Resurgence and recent advances

With the stated obstructions, bio-lubricants have undergone recent investigation wherein the major concerns have somewhat been overcome. Recent studies are noted and concluded in this section. For instance, Sharma *et al* (2015) evaluated the performance of canola oil and biodiesel based lubricant to develop bio-lubricants and reported kinematic viscosities of 12 and 670 cSt, respectively [51]. Another investigation by Jeevan and Jayaram (2018) revealed that the biodiesel developed from neem and mahua has potentially lower iodine values that suggest their higher oxidation stability. Also, the developed bio-lubricants have a high VI of 194 and 208 for neem and mahua, respectively [8]. In the exact string, Sneha and Muhammad (2019) used rice bran, Karanja, and jatropha oil as base feedstocks for industrial-scale bio-lubricant, evaluating pour points of −8 °C, −6 °C, and −7.5 °C, respectively, and iodine values of 93, 75, and 91 respectively [24]. Recently, Almeida *et al* (2020) experimented with cardanol-based bio-lubricants revealing the viscosity of 7000 cP [5]. Such outcomes from the novel materials depict a potentially strong viscosity of bio-lubricants. Nogales-Delgado *et al* (2020) synthesized the bio-lubricant from rapeseed oil and castor oil, evaluating viscosity of 10 and 2018 cSt, respectively [12]. However, newer investigations include synthesizing bio-lubricants from waste cooking oil incorporating ionic liquids and lipase, revealing conversion yield of 81% and selectivity of 82%. Moreover, the synthesized system has a high VI of 149 [52]. Therefore, from the author's point of view tailoring the properties of base feedstocks, incorporating the chemical species, focusing on the feedstock base material, etc could help in overcoming the existing limitations of bio-lubricants [7].

1.9 Conclusion

This chapter focuses on briefing the lubricant's characteristics and then stating the bio-lubricants concept. Some of the commonly investigated lubricating oils like castor oil, rapeseed oil, sunflower oil, etc, were delineated with their practical purpose. The evolution of bio-lubricants and the proliferation with their advantages were examined.

The focus has been given to the synthesis, tribo-chemical behaviour, and structure effect on friction/wear and load-bearing capacity of various bio-lubricating oils used nowadays. The literature has inferred a tremendous demand for vegetable oils. Properties of bio-lubricants have been explained with tribo-chemical behaviour. The reaction mechanism from initiation to decomposition during the synthesis has been summarized with a plausible pathway. Resurgence in the field of bio-lubricants has been discussed stating recent investigations. Additionally, technical issues are highlighted, and the possible ways of overcoming them have been suggested with the support of advances. Lastly, it was concluded that bio-lubricants sectors are the broader area for industrial-scale application, which is yet to be examined and can expand its future to the commercial level.

References

[1] Soni S and Agarwal M 2014 Lubricants from renewable energy sources—a review *Green Chem. Lett. Rev.* **7** 359–82

[2] Masripan N A, Salim M A, Omar G, Mansor M R, Saad A M, Hamid N A, Syakir M I and Dai F 2020 Vegetable oil as bio-lubricant and natural additive in lubrication: a review *Int. J. Nanoelectron. Mater.* **13** 161–76

[3] Reeves C J, Siddaiah A and Menezes P L 2017 A review on the science and technology of natural and synthetic biolubricants *J. Bio-Tribo-Corros.* **3** 1–27

[4] Tulashie S K and Kotoka F 2020 The potential of castor, palm kernel, and coconut oils as biolubricant base oil via chemical modification and formulation *Therm. Sci. Eng. Prog.* **16** 100480

[5] De Almeida M O, Silva L R R, Kotzebue L R V, Maia F J N, Acero J S R, Mele G, Mazzetto S E, Sinatora A and Lomonaco D 2020 Development of fully bio-based lubricants from agro-industrial residues under environmentally friendly processes *Eur. J. Lipid Sci. Technol.* **122** 1900424

[6] Soufi M D, Ghobadian B, Mousavi S M, Najafi G and Aubin J 2019 Valorization of waste cooking oil based biodiesel for biolubricant production in a vertical pulsed column: Energy efficient process approach *Energy* **189** 116266

[7] Cecilia J A, Plata D B, Saboya R M A, de Luna F M T, Cavalcante C L and Rodríguez-Castellón E 2020 An overview of the biolubricant production process: challenges and future perspectives *Processes* **8** 257

[8] Puttaswamy J T and Ramachandra J S 2018 Experimental investigation on the performance of vegetable oil based cutting fluids in drilling AISI 304L using taguchi technique *Tribol. Online.* **13** 50–6

[9] Mehdi S, Asghari A, Ghobadian B and Soufi M D 2019 Conversion of *Pistacia atlantica mutica* oil to trimethylolpropane fatty acid triester as a sustainable lubricant *Biomass Convers. Biorefin.* **10** 139–48

[10] Taha-Tijerina J, Aviña K and Diabb J M 2019 Tribological and thermal transport performance of SiO_2-based natural lubricants *Lubricants* **7** 71

[11] Chen J *et al* 2019 From ethyl biodiesel to biolubricants: options for an Indian mustard integrated biorefinery toward a green and circular economy *Ind. Crop. Prod.* **137** 597–614

[12] Encinar J M, Nogales-Delgado S, Sánchez N and González J F 2020 Biolubricants from rapeseed and castor oil transesterification by using titanium isopropoxide as a catalyst: production and characterization *Catalysts.* **10** 366

[13] Wahyuningsih T D and Kurniawan Y S 2020 Synthesis of dioxo-dioxane and dioxo-dioxepane ethyl oleate derivatives as bio-lubricant base stocks *Indones. J. Chem.* **20** 503–9

[14] Muhammad C 2018 Biodiesel production from *Ceiba pentandra* seed oil using CaO derived from snail shell as catalyst *Pet. Sci. Eng.* **2** 7

[15] Soni S and Agarwal M 2014 Lubricants from renewable energy sources—a review *Green Chem. Lett. Rev.* **7** 359–82

[16] Arumugam S, Sriram G and Subadhra L 2012 Synthesis, chemical modification and tribological evaluation of plant oil as bio-degradable low temperature lubricant *Proc. Eng.* **38** 1508–17

[17] Lathi P S and Mattiasson B 2007 Green approach for the preparation of biodegradable lubricant base stock from epoxidized vegetable oil *Appl. Catal.* B **69** 207–12
[18] de Araujo L R R, Scofield C F, Pastura N M R and Gonzalez W de A 2006 H_3PO_4/Al_2O_3 catalysts: characterization and catalytic evaluation of oleic acid conversion to biofuels and biolubricant *Mater. Res.* **9** 181–4
[19] Quinchia L A, Delgado M A, Valencia C, Franco J M and Gallegos C 2009 Viscosity modification of high-oleic sunflower oil with polymeric additives for the design of new biolubricant formulations *Environ. Sci. Technol.* **43** 2060–5
[20] Quinchia L A, Delgado M A, Valencia C, Franco J M and Gallegos C 2010 Viscosity modification of different vegetable oils with EVA copolymer for lubricant applications *Ind. Crop. Prod.* **32** 607–12
[21] Sharma R V and Dalai A K 2013 Synthesis of bio-lubricant from epoxy canola oil using sulfated Ti-SBA-15 catalyst *Appl. Catal. B: Environ.* **142–143** 604–14
[22] Tan S G and Chow W S 2010 Biobased epoxidized vegetable oils and its greener epoxy blends: a review *Polym.—Plast. Technol. Eng.* **49** 1581–90
[23] Zhang W, Wu J, Yu S, Shen Y, Wu Y, Chen B, Nie K and Zhang X 2020 Modification and synthesis of low pour point plant-based lubricants with ionic liquid catalysis *Renew. Energy* **153** 1320–29
[24] Sneha E, Rani S and Arif M 2019 Evaluation of lubricant properties of vegetable oils as base oil for industrial lubricant *IOP Conf. Ser.: Mater. Sci. Eng.* **624** 012022
[25] Hwang H S and Erhan S Z 2001 Modification of epoxidized soybean oil for lubricant formulations with improved oxidative stability and low pour point *J. Am. Oil Chem. Soc.* **78** 1179–84
[26] Zainal N A, Zulkifli N W M, Gulzar M and Masjuki H H 2018 A review on the chemistry, production, and technological potential of bio-based lubricants *Renew. Sustain. Energy Rev.* **82** 80–102
[27] Ravasio N, Zaccheria F, Gargano M, Recchia S, Fusi A, Poli N and Psaro R 2002 Environmental friendly lubricants through selective hydrogenation of rapeseed oil over supported copper catalysts *Appl. Catal.* A **233** 1–6
[28] Kržan B and Vižintin J 2003 Tribological properties of an environmentally adopted universal tractor transmission oil based on vegetable oil *Tribol. Int.* **36** 827–33
[29] Gryglewicz S, Piechocki W and Gryglewicz G 2003 Preparation of polyol esters based on vegetable and animal fats *Biores. Technol.* **87** 35–9
[30] Anastopoulos G, Zannikou Y, Stournas S and Kalligeros S 2009 Transesterification of vegetable oils with ethanol and characterization of the key fuel properties of ethyl esters *Energies* **2** 362–76
[31] Doll K M and Erhan S Z 2005 Synthesis of carbonated fatty methyl esters using supercritical carbon dioxide *J. Agric. Food Chem.* **53** 9608–14
[32] Widayat W, Wibowo A D K and Hadiyanto H 2013 Study on production process of biodiesel from rubber seed (hevea brasiliensis) by *in situ* (trans)esterification method with acid catalyst *Energy Proc.* **32** 64–73
[33] Sripada P K, Sharma R V and Dalai A K 2013 Comparative study of tribological properties of trimethylolpropane-based biolubricants derived from methyl oleate and canola biodiesel *Ind. Crops Prod.* **50** 95–103
[34] Nohair B, Especel C, Marécot P, Montassier C, Hoang L C and Barbier J 2004 Selective hydrogenation of sunflower oil over supported precious metals *C. R. Chim.* **7** 113–8

[35] Freedman B, Butterfield R O and Pryde E H 1986 Fast capillary chromatography, spectra-physics *J. Am. Oil Chem. Soc.* **63** 1375
[36] Shomchoam B and Yoosuk B 2014 Eco-friendly lubricant by partial hydrogenation of palm oil over Pd/γ-Al_2O_3 catalyst *Ind. Crops Prod.* **62** 395–99
[37] Masudi A and Muraza O 2018 Vegetable oil to biolubricants : review on advanced porous catalysts vegetable oil to biolubricants : review on advanced porous 2 catalysts *Energy Fuels* **32** 10295–310
[38] Choo H P, Liew K Y, Liu H F and Seng C E 2001 Hydrogenation of palm olein catalyzed by polymer stabilized Pt colloids *J. Mol. Catal.* A **165** 127–34
[39] Isbell T A, Edgcomb M R and Lowery B A 2001 Physical properties of estolides and their ester derivatives *Ind. Crops Prod.* **13** 11–20
[40] Chen Y, Biresaw G, Cermak S C, Isbell T A, Ngo H L, Chen L and Durham A L 2020 Fatty acid estolides: a review *J. Am. Oil Chem. Soc.* **97** 231–41
[41] Wagner H, Luther R and Mang T 2001 Lubricant base fluids based on renewable raw materials: Their catalytic manufacture and modification *Appl. Catal.* A **221** 429–42
[42] Karimi S, Sadjadi S, Bahri-Laleh N and Nekoomanesh-Haghighi M 2021 New less-toxic halloysite-supported ionic liquid/$AlCl_3$ oligomerization catalysts: a comparative study on the effects of various ionic liquids on the properties of polyalphaolefins *Mol. Catal.* **509** 111648
[43] Saurabh T, Patnaik M, Bhagst S L and Renge V 2011 Epoxidation of vegetable oils: a review *Int. J. Adv. Eng. Technol.* E **2** 459–501
[44] Wu X, Zhang X, Yang S, Chen H and Wang D 2000 The study of epoxidized rapeseed oil used as a potential biodegradable lubricant *J. Am. Oil Chem. Soc.* 561–63
[45] McNutt J and He Q S 2016 Development of biolubricants from vegetable oils via chemical modification *J. Ind. Eng. Chem.* **36** 1–12
[46] Panchal T M, Patel A, Chauhan D D, Thomas M and Patel J V 2017 A methodological review on bio-lubricants from vegetable oil based resources *Renew. Sustain. Energy Rev.* **70** 65–70
[47] Azahar W N A W, Jaya R P, Hainin M R, Bujang M and Ngadi N 2016 Chemical modification of waste cooking oil to improve the physical and rheological properties of Asphalt binder *Constr. Build. Mater.* **126** 218–26
[48] Lligadas G, Ronda J C, Galià M and Cádiz V 2013 Renewable polymeric materials from vegetable oils: a perspective *Mater. Today* **16** 337–43
[49] Schäffner B *et al* 2014 Synthesis and application of carbonated fatty acid esters from carbon dioxide including a life cycle analysis *ChemSusChem* **7** 1133–39
[50] Owuna F J, Dabai M U, Sokoto M A, Dangoggo S M, Bagudo B U, Birnin-Yauri U A, Hassan L G, Sada I, Abubakar A L and Jibrin M S 2020 Chemical modification of vegetable oils for the production of biolubricants using trimethylolpropane: a review *Egypt. J. Pet.* **29** 75–82
[51] Sharma R V, Somidi A K R and Dalai A K 2015 Preparation and properties evaluation of biolubricants derived from canola oil and canola biodiesel *J. Agric. Food Chem.* **63** 3235–42
[52] Zhang W, Ji H, Song Y, Ma S, Xiong W, Chen C, Chen B and Zhang X 2020 Green preparation of branched biolubricant by chemically modifying waste cooking oil with lipase and ionic liquid *J. Clean. Prod.* **274** 122918
[53] Cai S and Wang L 2011 Epoxidation of unsaturated fatty acid methyl esters in the presence of SO_3H-functional Brønsted acidic ionic liquid as catalyst *Chin. J. Chem. Eng.* **19** 57–63

IOP Publishing

Multifunctional Bio-Based Lubricants
Synthesis, properties and applications
Ashish K Kasar, Arpith Siddaiah and Pradeep L Menezes

Chapter 2

Bio-based lubricants and their environmental sustainability

Satish Vasu Kailas, Jeng Haur Horng and P S Suvin

The evolution of lubricants from petroleum-based products has resulted in significant changes in the current and expanding manufacturing industry. Mineral-based oils, which are toxic, non-biodegradable, and unsustainable, are used in the majority of lubricants. A major issue is their improper disposal, which causes contamination of surface water and groundwater, as well as agricultural products and food. Thus, the development of bio-lubricants that meet all the criteria of a 'closed loop' product is essential. The requirements that need to be fulfilled are: (i) the ingredients chosen should be completely non-toxic (this is taken as the most important criteria since the unsafe disposal practice cannot be easily stopped); (ii) the rate of consumption or usage or the raw materials that go into making the lubricant should be less than or at least match the rate of generation or replenishment; (iii) the discard should be biodegradable. Using this philosophy an attempt is made to make sustainable cutting oil emulsions. A green cutting fluid (GCF) is made from a vegetable oil with food-grade emulsifiers. The additives, which form an important component of the formulation, and generally the toxic part of the formulation, are also plant derived and non-toxic. The cutting oil emulsion is then tested using standard tests procedures for emulsion stability, bacterial growth, corrosion resistance and cutting efficiency. GCF fulfils many of the characteristics of commercial formulations while still being environmentally friendly. Nevertheless, GCF formulation with non-toxic emulsifiers and natural additives is a good basis for further development and use of non-toxic tribological products.

2.1 Introduction

Lubricant is a material used to minimize friction and wear between moving surfaces. Lubricants are classified into liquid, solid, semi-solid, and gaseous. Most of the lubricants, although not all, are liquids. Mineral oil is used as a base in conventional lubricants, coupled with various additives for performance enhancement, which is harmful and unsustainable. Mineral oil is obtained from fossil fuels, which are one of our planet's scarce resources. As a result, adopting sustainable product development technology is essential. Mineral oil has low biodegradability, making it a long-term source of pollution in the environment. Mineral oil-based compositions can pollute groundwater for decades, stunting plant development and shortening the life span of aquatic life. Liquid lubricant designed for machining and manufacturing processes, such as drilling, milling, turning, reaming, boring, sawing, parting and stamping, is known as cutting fluid (CF) or metalworking fluid (MWF). It is made up of a mixture of oil, emulsifiers, and additives that are largely produced from petroleum. In the manufacturing industry, metal cutting fluids or coolants are used to keep the work zone at a consistent temperature, lubricate the chip-tool contact, and flush away chips [1]. There are several economic advantages of using cutting fluids, such as: (i) reduction in tool costs as tool life increases by reducing tool wear; (ii) increase in production rate as cutting fluids reduce heat and friction between tool and workpiece, enabling higher speed of cutting; (iii) avoiding built-up edge (BUE) formation, thus reducing downtime and regrinding of tools, which reduces labour and power costs. There are various categories of cutting fluids, which include neat oils, synthetic cutting oils, semi-synthetic cutting oils, oil-in-water emulsions, water-in-oil emulsions, aerosols (mists), pastes and gels. They may be obtained from distillates of petroleum, animal fats, plant oils, water and air, or other raw ingredients. Depending on the context and the type of cutting fluid being considered, it may be referred to as metal cutting fluid, cutting fluid, MWF, cutting oil, cutting compound, coolant, or lubricant. Consumption of MWFs by North American manufacturers was over two billion gallons in 2002 [2]. Due to their advantages, the consumption of cutting fluids is increasing in the machining industry. The compositions of MWFs became more complex as the cutting processes became more difficult. To satisfy the performance expectations of industrial industries, cutting fluids are currently strengthened with a combination of oil, water, emulsifiers, and chemical additives [3]. The United States uses 100 million gallons of metalworking oil per year, according to estimates [4].

Manufacturing operations that use conventional cutting fluids are generally not very eco-friendly, and they produce a lot of goods and by-products that are damaging to employees and the environment. A cutting fluid also contains a variety of additives, such as mineral oil and emulsifiers, to improve its performance. These additives are extremely harmful to the environment, as well as non-renewable and non-biodegradable. This non-biodegradable nature causes serious soil and water pollution. It also leads to biomagnification. Worldwide annual consumption of these MWFs is increasing; more than 50% of this is lost or disposed off in an unregulated manner into the environment causing environmental and health hazards [5]. Apart from the environmental hazards, they also pose significant health issues throughout their life cycle. The toxicity of the commercial fluid enters the food chain, which is very dangerous not only for the present generation but also for future generations. Many complex mixtures of biocides used in cutting oil make it an irritant and the reason for several occupational diseases like dermatitis, oil acne, respiratory-tract infections, life threatening diseases like skin cancer, respiratory infection etc and many other allergic reactions [6]. Health hazards from MWFs are due to the chemical or physical nature of their formulations or from the direct or indirect activities of microorganisms.

Cutting fluid alkalinity facilitates the selective growth of many harmful bacteria. After a period of immobility, many incompletely oxidised products of anaerobic microbes produce a bad odour. Cutting fluid microbiology is extremely complicated, with different bacteria prevailing at different stages of the process, each with a particular metabolic activity. The microbial toxins and enzymes produced as a result, especially in water-soluble cutting fluids, as well as the often employed chemical biocides to kill bacteria, are carcinogenic. Cutting fluids' metal particle contamination and poisonous by-products impair the quality of cutting fluids after they've been used. As a result, minimising fluid disposal is a major concern because it is both costly and hazardous to the environment. Studies and literature on harmful effects of the commercial cutting fluids show that the concentration of chemicals in cutting fluid is shooting up. The disastrous impact of commercial cutting fluid on living things and the environment is represented in figure 2.1.

Pollution and environmental health aspects have become immensely important public issues in recent years. Use of eco-friendly MWFs is the possible answer to minimising adverse environmental effects and protecting ecology. As a result, it is critical to create an environmentally acceptable, non-toxic, and long-lasting product with good cutting fluid capabilities. To overcome the limits of mineral oil-based cutting fluids, scientists and tribologists have devised a variety of alternatives. These alternative solutions include synthetic lubricants, animal tallow and solid lubricants. Due to its renewable, environmentally friendly, relatively non-toxic, and rapidly biodegradable attributes, vegetable oil is another highly appealing mineral oil substitute. Vegetable oils have the disadvantages of low thermal and oxidation stability. Vegetable oils are esters of glycerol and fatty acids. Advantages of vegetable oils include good solvency, high lubricity, low volatility, and high load carrying capacity, low emission of hydrocarbons, higher fire resistance and good thermal properties [7, 8]. Eco-friendly/biodegradable lubricants are predicted to have a global volume share of roughly 15% in the

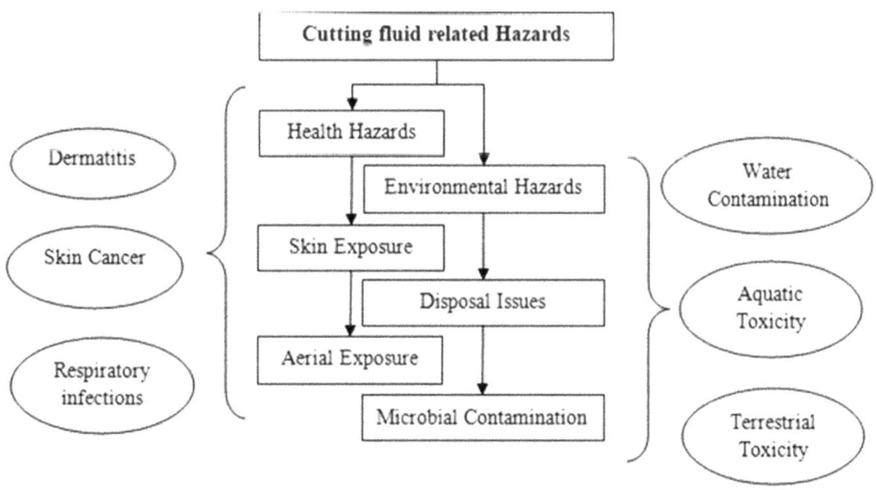

Figure 2.1. Issues of cutting fluid. Copyright Dr P S Suvin.

next 10–15 years, with up to 30% in some countries [9]. Vegetable oil modifications have also been demonstrated to boost results in a number of investigations. In some machining situations, vegetable oil has the potential to degrade. Because viscosity is higher, a protective layer forms between the mating surfaces, preventing lubricant breakdown [10]. Vegetable oils have a number of drawbacks, including low thermal and oxidative stability, high freezing temperatures, and poor corrosion protection [11]. The bulk of vegetable oils have four to twelve different fatty acids in them. Weather and geoclimate play a major role in determining the amount of individual fatty acids [12].

2.2 Closed loop concept of sustainability

We all rely on the Earth to live, with little regard for the impact it has on others. Choices between needs and wants or desires are frequently at odds. We are depleting the Earth's resources at a rate that will leave little for future generations, resulting in a terrible downward cycle in many regions of the planet. Environmental stress is being caused by increasing demand for scarce resources and rising living standards, which might be addressed through sustainable development. For any product, each stage of life cycle has subsidiary chains, cause and effect loops which are very difficult to comprehend. Many approaches started adding to the basic framework defined for sustainability, such as integrating environmental requirements in every stage of product cycle [13]. Further, many frameworks were made on sustainable manufacturing, even frameworks involving ethics and accountability, as shown in figure 2.2.

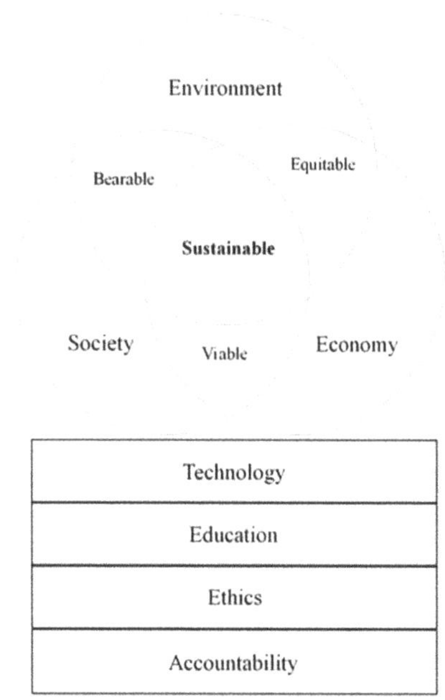

Figure 2.2. Framework of sustainability. Copyright Dr P S Suvin.

Sustainability of constituent materials plays an imperative part in maintaining the integrity of ecology. Sustainability concepts of lubricant materials have two sides. The first consideration is the reserve's source or origin: petroleum reserves are declining rapidly, whereas bio-reserves are renewable. The second point to consider is the CO_2 cycle; for oleo compounds, the CO_2 released is equal to the CO_2 taken up by the plants, closing the loop. In contrast, the CO_2 cycle for petroleum-derived goods is not closed but open, resulting in an increase in atmospheric CO_2 levels and so contributing to global warming, in addition to the impacts of exposing petroleum derivatives to the environment [14].

Schematic representations of open and closed loop systems are shown in figures 2.3 and 2.4, where the discard is the reason for making the cycle open, further leading to aggregation of hazardous material and environmental stress. For

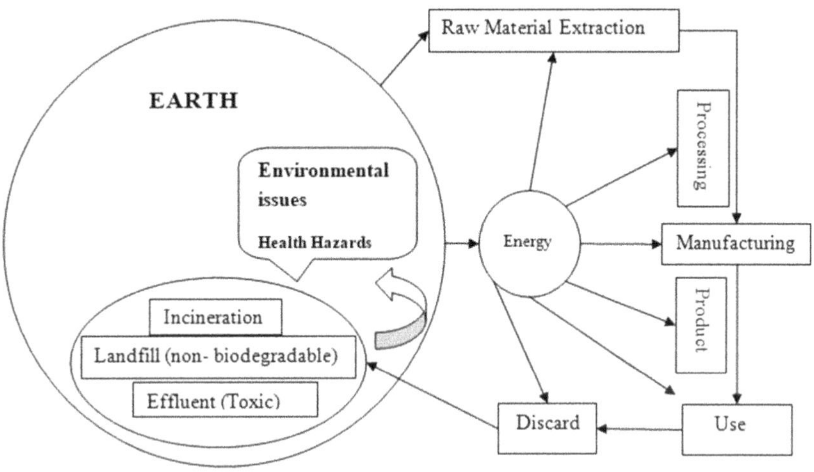

Figure 2.3. Open loop concept.

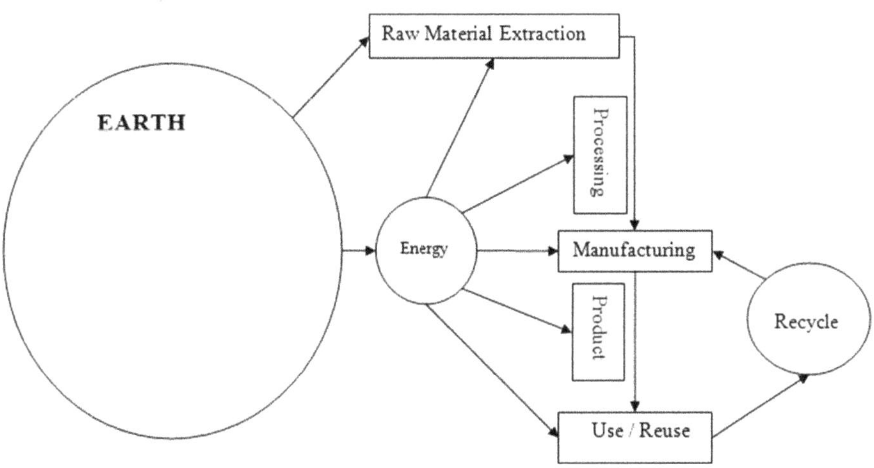

Figure 2.4. Closed loop concept. Copyright Dr P S Suvin.

a sustainable cycle, the rate of product consumption or usage should be less than or at least match the rate of regeneration or replenishment. The current consumption rate of mineral reserves is not comparable with its replenishment rate, which means it is not sustainable if we continue to use the petroleum reserves at the same rate.

Sustainability is based on the basic principle of maintaining productive harmony, fulfilling the social, economic and environment responsibilities for the well-being of present and future generations [13]. Conventional mineral-based lubricants produced from non-renewable sources will use up the mineral deposits faster than their generation and leave a big carbon footprint. Thus, it becomes an open loop generating waste harmful to the environment. But in the case of vegetable oil-based lubricant, there is balance between production and generation. The waste generated is non-toxic and can be disposed of easily into the environment and can also be re-used for a different application, thus, 'closing the loop' without leaving any carbon footprint and causing no harm to the environment.

As a result of extensive use of mineral oil-based lubricants, toxins are getting accumulated and finally affecting the system balance. In this backdrop, vegetable oil- based cutting fluids have gained attention over the last many decades. Figure 2.5 represents a flowchart for checking the sustainability aspect where RR and ER represent replenishment rate and extraction rate, respectively.

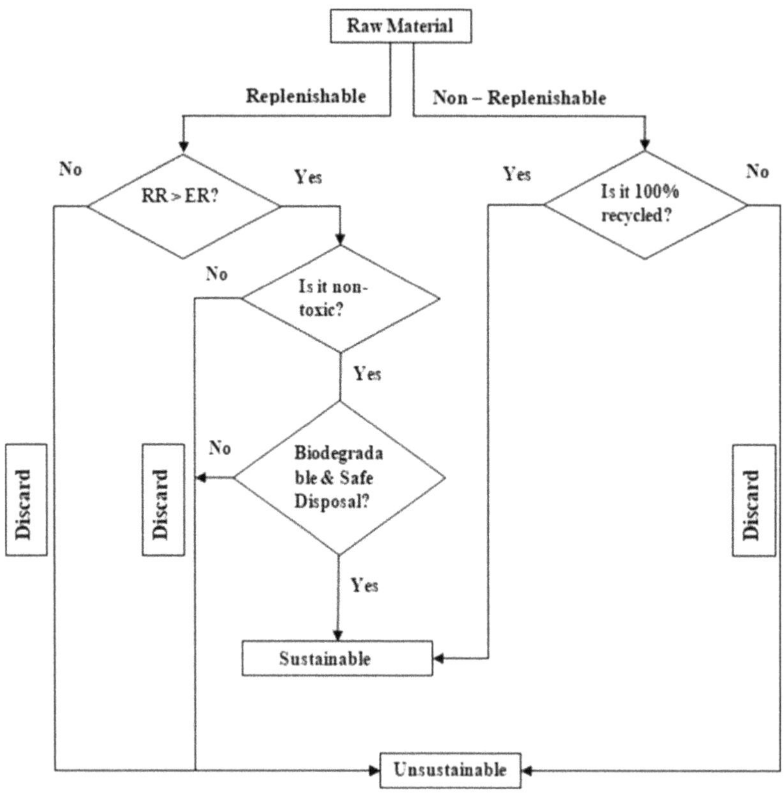

Figure 2.5. Representation of closed loop concept of sustainability. Copyright Dr P S Suvin.

A product undergoes three life stages during its life cycle: manufacturing, use and end of life/disposal. For any product to be sustainable, the life cycle of the product needs to be a 'closed loop'. In the closed loop philosophy, simple questions are asked regarding the raw materials used for making the product and on what happens to the product at its end of life. In the extraction stage one question needs to be asked: is the raw material replenished by nature? If the raw material is replenishable then the question to be asked is on the extraction and replenishment rate of the raw material. If the extraction rate is greater than the replenishment rate then the cycle is unsustainable. The extraction rate needs to be lower than the replenishment rate for the cycle to be sustainable. If the raw material is something that is not replenished, if it is discarded after use, then the cycle is not sustainable. The cycle that uses materials that cannot be replenished will be sustainable only if the raw material is completely recycled. Then come the conditions to be met during disposal, given that only materials that are replenished can be disposed. The critical question during disposal is on the toxicity and biodegradability. The disposed of product should not be toxic and should be biodegradable. Only if the product meets these criteria and follows a 'closed loop' will it be sustainable.

For example: given that cutting fluids use mineral oils that are being extracted at a rate faster than what nature can replenish and given that the product, which is highly toxic and has limited biodegradability, is generally discarded into the environment makes it highly unsustainable. Thus, the development of cutting oil that meets all the criteria of a 'closed loop' product is essential. The requirements that need to be fulfilled are:

(i) The ingredients chosen should be completely non-toxic (this is taken as the most important criteria since the unsafe disposal practice cannot be easily stopped).
(ii) The rate of its consumption or usage or the raw materials that go into making of the cutting oil should be less than or at least match the rate of generation or replenishment.
(iii) The discard should be biodegradable.

2.3 Synthesis and evaluation of green cutting fluid

Using this 'closed loop' approach a green cutting fluid (GCF) was prepared. To meet the requirement of the conditions for sustainability, a non-toxic vegetable oil is used as the raw material. Coconut oil is chosen as it has a high degree of saturation, making it more stable to oxidation (the 'bane' of vegetable oils) than other vegetable oils [14]. Coconut oil is scientifically known as Cocos nucifera. Coconut oil (extracted by crushing the dried kernel of coconut grown in coconut palms) supplied from an oil mill was used 'as received', as the base oil for the GCF. Hydrophilic–lipophilic balance (HLB) provides a rough guideline regarding the selection of emulsifying oils in water. HLB value of the base oil (coconut oil) was determined experimentally and emulsifiers to match this HLB value were selected. Commercially available non-toxic emulsifiers E1, E2, and E3 obtained from different emulsifier suppliers were used for this purpose. The other additives used to prepare the GCF included green additives extracted from the leaves of medicinal plants available in India, AzadirachtaIndica Oil (A1), Cymbopogon Citratus oil (A2),Ocimum Tenuiflorum oil (A3), and jaggery syrup (A4). A blend was formed from these ingredients (coconut oil (40%w/w), E1 (31%w/w); E2 (8%w/w); E3 (11% w/w) and additives—A1 (2%w/w); A2 (3%w/w); A3 (2%w/w); A4 (3%w/w)). The additives were chosen based on their non-toxic, anti-corrosive, antibacterial, and skin viability properties. The blend was then diluted by mixing it with deionised water, in the ratio of 1:20. The properties of the synthesised GCF were evaluated as per standard test procedures (for emulsion stability, toxicity, anti-corrosion, anti-bacterial test and biodegradability test) and compared against commercially available cutting fluids (CCFs).

Particle size, zeta potential, and oven test results were used to determine the emulsion stability of GCF. A 90Plus Particle Size Analyzer was used to quantify particle size and zeta potential (Brookhaven Instruments Corporation). The average particle size of GCF was determined to be 220.1 nm, compared to 200 nm for CCFs, and its zeta potential was 47.46 mV, compared to 42.23 mV for CCF4. Storage stability (shelf life) of the emulsion was determined by inspecting the sample visually. On visual inspection after 45 days, GCF had the least separation or creamy layer, while only a few CCFs had a creamy layer with an oily phase. The samples' pH levels were also determined. The thermal stability of the emulsion sample was assessed using an oven test according to ASTM D3707-89, which measured the separation of the emulsion sample into oil and water after heating a 100 mL test sample at 85 °C for 48 h in a thermostatically controlled oven [16].

A few CCFs showed higher oil and water separation, whereas GCF showed the least amount of emulsion separation. The stability of GCF was shown to be superior to that of CCFs based on all of the metrics measured. There were five commercial cutting fluids—CCF1, CCF2, CCF3, CCF4, CCF5. GCF is numbered 6. CCF2 is synthetic, which gives a transparent colour. Characteristic properties of cutting fluids tested are shown in table 2.1.

Table 2.1. Characteristic properties of cutting fluids tested. Copyright Dr P S Suvin.

	Test samples					
Evaluation test	GCF	CCF1	CCF2	CCF3	CCF4	CCF5
Creamy layer, mL	0	0	0	2	0	1
Water separation, mL	0	0	0	0	2	0
Droplet size (Pz), nm	220.1	189.0	–	230	260	190
Zeta potential (mV)	47.46	49.10	–	39.90	42.23	35.63
Viscosity at 45 and 100 (mPa s)	95.94 and 10.90	66.0 and 8.45	70.43 and 14.89	39.49 and 5.36	27.96 and 4.54	21.25 and 3.81
Oven tests (70 °C–80 °C)	Stable	Stable	Stable	Not stable	Stable	Stable
pH	7.5	9.5	8.5	9.2	8.8	9
Odour	Lemon	Mineral oil	Nil	Mineral oil	Mineral oil	Mineral oil

2.3.1 Toxicity evaluation of constituents

2.3.1.1 Fish toxicity test OECD-203

The OECD 203 test technique for acute fish toxicity test was used to determine the aquatic toxicity of each constituent and the final composition of cutting fluid samples. The species are chosen based on certain crucial practical characteristics such as availability throughout the year, ease of maintenance, and testing convenience, among others [17]. The fish should be in good health and free from any malformation. Based on the practical criteria the fish species selected on recommendation for testing is *Brachydanio rerio* (zebra fish). Experiments were conducted on a pool of fish that had a mortality rate of less than 5% prior to testing. For 96 h, seven fish were exposed to test compounds. At the geometric series of 2.2, the concentration of the test material increased. Every 24 h, the dissolved oxygen and pH of each sample were measured. At 24, 48, 72, and 96 h, mortality was observed. The samples' lethal concentration 50 (LC50) was calculated by measuring the concentration at which 50% of the fish are killed.

The aquatic system is regarded as reasonably non-toxic to test samples with LC50 > 1000 mg L^{-1} of water. The acute fish toxicity of coconut oil, emulsifiers, various green additives, GCF, and CCFs was investigated. Coconut oil has an LC50 value of >2342.56 mg L^{-1}, while emulsifiers and additives have an LC50 value of >1064 mg L^{-1} [15, 16]. The LC50 of the GCF with green additions is 1064 mg L^{-1}.

CCF has an LC50 value of less than 100 mg L^{-1} (with the exception of CCF2, which has an LC50 of 800 mg L^{-1}) and is 10 times more hazardous than GCF. When 100 mg L^{-1} CCF, CCF3, CCF4, and CCF5 commercial cutting fluid was used in a zebra fish survival study, all of the fish died within 24 h, however, when 1064 mg L^{-1} GCF was used, no toxicity was found (table 2.2).

This finding shows that even at 10 times greater concentrations of GCF, there is no evidence of aquatic toxicity. According to OECD 203 test procedures, GCF outperforms CCF by being non-toxic at LC50 > 1000 mg L^{-1} and harmful at LC50 800 mg L^{-1}. Figure 2.6 illustrates the situation. GCF is non-toxic and helps contribute to green and sustainable manufacturing processes without polluting or endangering the environment. It is made entirely of renewable resources.

We can clearly see that GCF's toxicity values are substantially lower than commercially available cutting fluids based on the OECD 203 Acute Fish Toxicity Tests (CCF).

Table 2.2. Observation results of LC50 in fresh samples [15], copyright 2021 SAGE Publications.

Samples	LC50 concentration of sample (mg L^{-1})
CCF1	<40
CCF2	<800
CCF3	<50
CCF4	<80
CCF5	<100
GCF 6	⩾1064

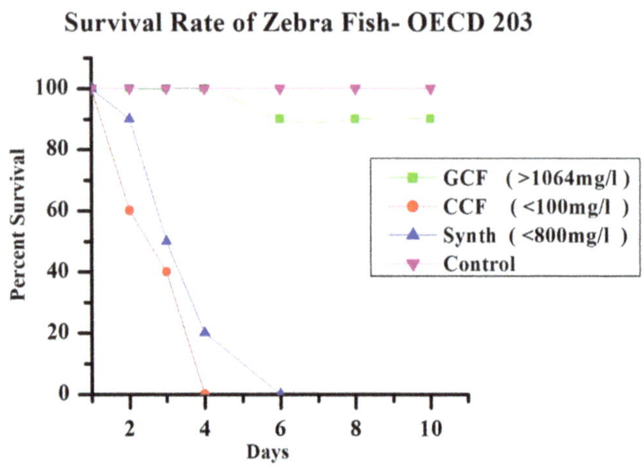

Figure 2.6. Observation results of LC50 of samples [15], copyright 2021 SAGE Publications.

Table 2.3. Test concentration for the sample.

Control	Control	1% (20 μL in 2 mL)	1% (20 μL in 2 mL)	2% (40 μL in 2 mL)	2% (40 μL in 2 mL)
4% (80 μL in 2 mL)	4% (80 μL in 2 mL)	6% (120 μL in 2 mL)	6% (120 μL in 2 mL)	10% (200 μL in 2 mL)	10% (200 μL in 2 mL)

Even at lower quantities, fish were unable to live in the system using commercial samples; this could be due to toxins present in the cutting fluids. The elements that make up the final composition of GCF, on the other hand, are non-toxic and environmentally benign, which could explain the rise in the LC50 value. Toxicity was examined using fish toxicity tests (OECD 203 standards), and it was discovered that GCF outperformed CCF by being non-toxic at LC50 > 1000 mg L^{-1} and toxic at LC50 800 mg L^{-1} [15].

2.3.1.2 Fish embryo test: FET-236

To test the effects of toxicity, zebra fish embryos were treated with different doses of solutions and water as a control at 6 h post fertilisation (hpf). Cutting fluids were used to treat zebra fish embryos at five different doses (1, 2, 4, 6, and 10). As stated in table 2.3, the concentration ranges were chosen based on the findings of the range-finding experiment, which identified a concentration that caused mortality and morphological deformities. The 6 hpf time point was chosen because it is easy to identify unfertilized eggs that enter the test by accident at this stage. Two embryos were placed in each well of a 24-well plate with a final volume of 2 mL per well and incubated at 26 °C with a 12:12 h light dark cycle. For each concentration test fluid and control, three replicates were done. To check for lethality and morphological defects, zebra fish embryos were inspected through a stereomicroscope (Olympus SZ51 stereomicroscope). Figure 2.7 and table 2.4 show the results of the embryo development observations.

Figures 2.8 and 2.9 represents the results obtained in fish embryo test with different concentrations. Copyright Dr P S Suvin.

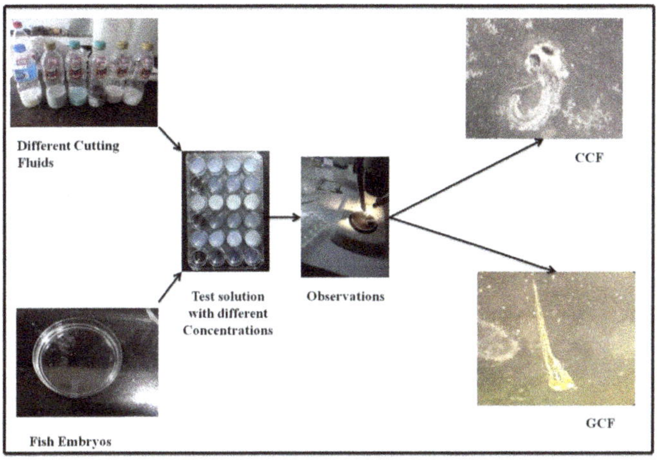

Figure 2.7. Fish embryo test procedure and observation results of LC50 in samples. Copyright Dr P S Suvin.

Table 2.4. Observations of embryo development. Copyright Dr P S Suvin.

Dead Embryo- Fully white Chorion (Outer layer of embryo)- In commercial cutting fluid 5 after 24 hr.		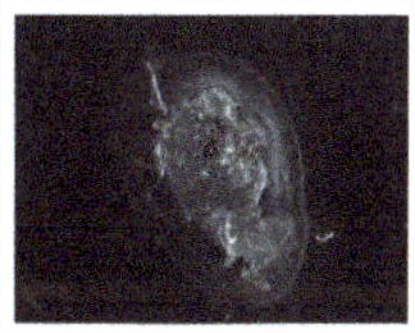
Alive Embryo- Transparent layer and chorion- In GCF after 48 hr	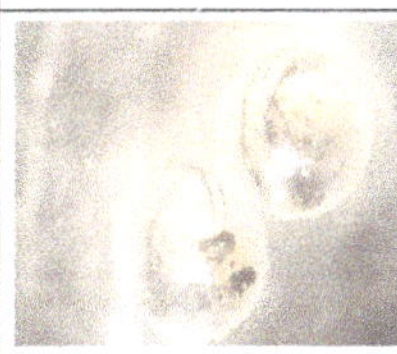	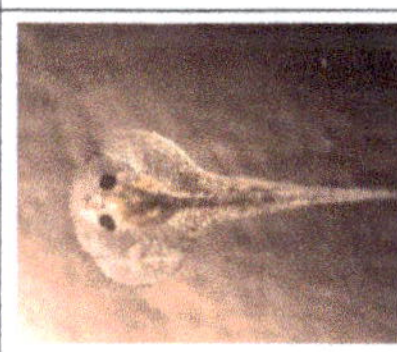
Dead Larvae– In commercial cutting fluid [CCF]	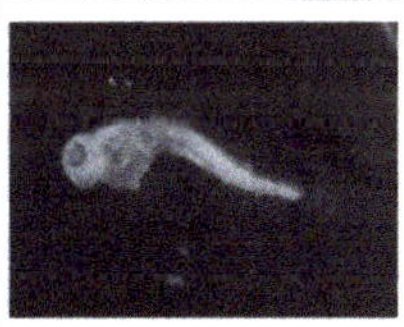	
Alive Larvae – In GCF	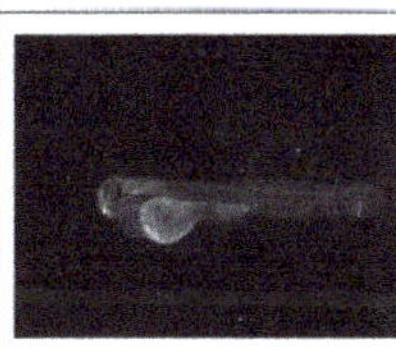	
Alive Larvae – In Control	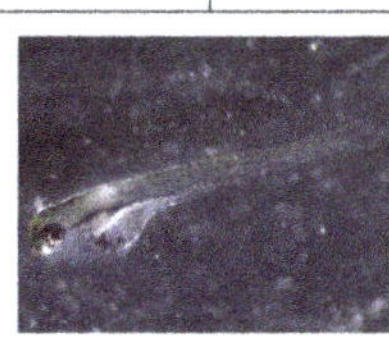	

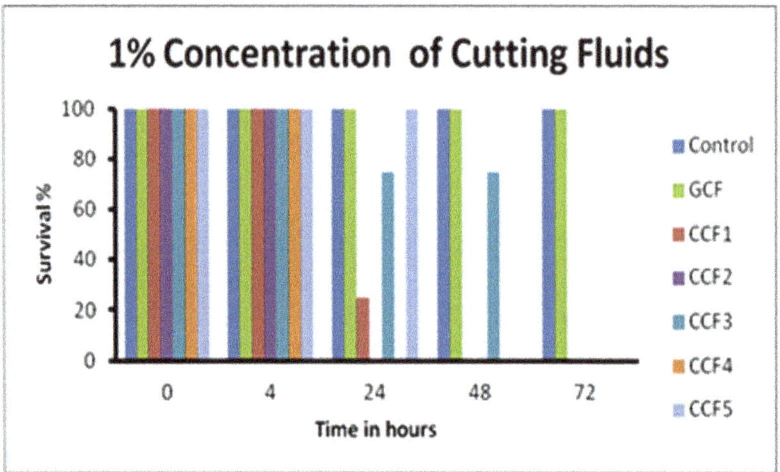

Figure 2.8. Observation results of LC50 in samples FET with 1% concentration of cutting fluid. Copyright Dr P S Suvin.

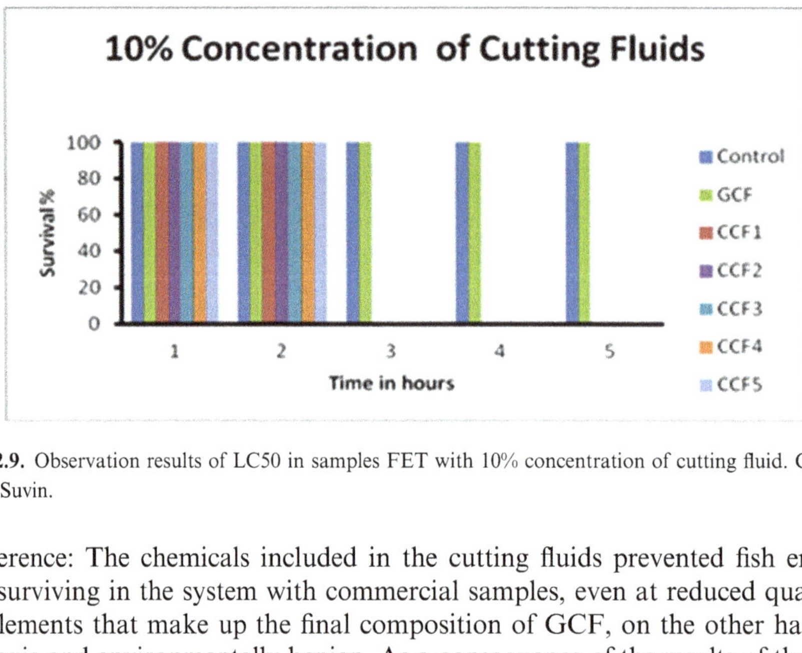

Figure 2.9. Observation results of LC50 in samples FET with 10% concentration of cutting fluid. Copyright Dr P S Suvin.

Inference: The chemicals included in the cutting fluids prevented fish embryos from surviving in the system with commercial samples, even at reduced quantities. The elements that make up the final composition of GCF, on the other hand, are non-toxic and environmentally benign. As a consequence of the results of the Acute Fish Toxicity Test and the Fish Embryo Test, we can conclude that the number of dangerous chemicals found in the GCF is low compared to the CCFs.

2.3.1.3 Skin cell viability test

Human keratinocyte cell lines (HaCaT) were employed as a model system to investigate the toxicity of cutting fluids. MTT (dimethylthiazol-diphenyl tetrazolium bromide, a yellow dye) cytotoxicity tests: The MTT test was used to determine the cytotoxicity of

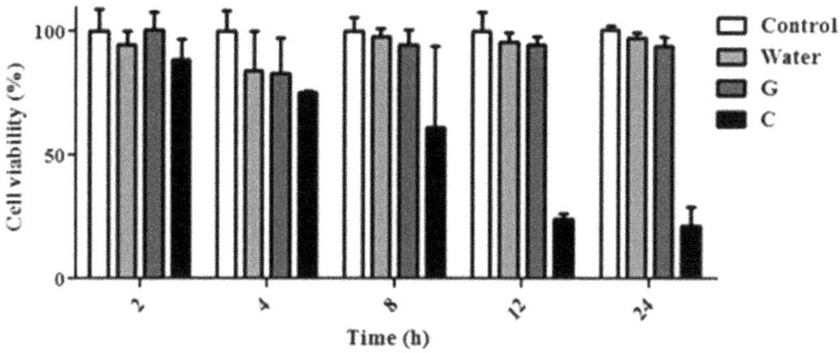

Figure 2.10. Cell line viability after treatment with cutting fluid samples. G—Green cutting fluid; C—Commercial mineral oil-based cutting fluid [16], copyright 2016 by Japanese Society of Tribologists.

cutting fluid samples *in vitro*. HaCaT cell lines (immortal human keratinocyte line) were grown in Dulbecco's Modified Eagle's Medium (DMEM, Sigma) supplemented with 10% foetal bovine serum and kept at 37 °C in 5% CO_2. MTT assay was used to analyse the cytotoxicity of cutting fluid samples (GCF, CCF) after incubation with 5104 cells in a 96-well plate at different time intervals ranging from 2 to 24 h. The cells were treated with 50 g mL^{-1} GCF and CCF for varying amounts of time. Each well was filled with 20 1 MTT dye (5 mg mL^{-1}) and incubated at 37 °C for 4 h. Depending on the vitality of the cells, MTT is converted to insoluble formazan crystals.

Replace the medium with an equal volume of dimethyl sulfoxide (DMSO) and set aside for 30 min to dissolve the formazan into a purple-colored solution. A SPECTRAmax PLUS 384 spectrophotometer was used to measure the absorbance at 550 nm. At different time points, cell viability was determined as a percentage of control cells (untreated cells) [16].

At 50 g mL^{-1}, the CCF sample caused considerable cell death, but the GCF sample (50 g mL^{-1}) had no effect on the cell lines' viability. The results (figure 2.10) show that CCF is toxic when compared to GCF, which has no harmful impact at a concentration of 50 g mL^{-1}. The viability of CCF cells was significantly reduced, whereas the viability of GCF cells was not significantly reduced. These findings show that CCFs are hazardous to human keratinocyte cell lines [16].

2.3.2 GC–MS analysis of commercial and GCF samples

Gas chromatography–mass spectroscopy (GC–MS) is a very compatible technique and the most commonly used technique for the identification and quantification purpose. GC–MS analysis will help in analysing the compounds present in CCFs [18]. Results are given in figure 2.11 and observations are shown in tables 2.5 and 2.6. The peaks of the GC–MS result are manually analyzed based on their retention time and probability to identify the components in the sample.

GC–MS was carried out on both GCF and CCF and it was found that CCF formulation consists of hazardous chemicals which cause respiratory damage as well as aquatic pollution. Five commercial samples and GCF were tested.

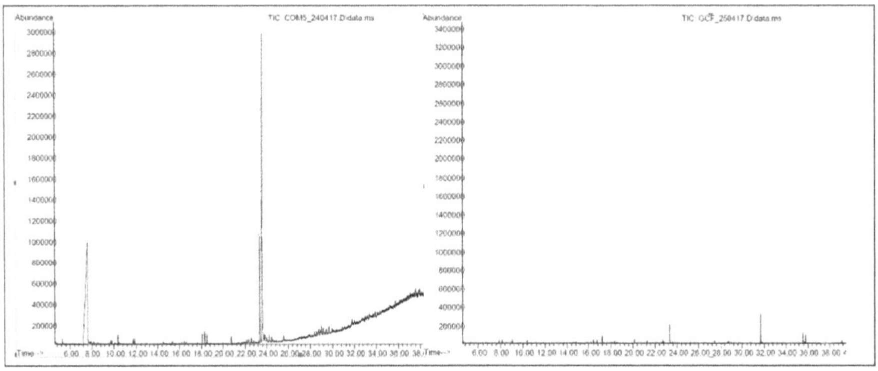

Figure 2.11. Compound peaks identified for GCF. Copyright Dr P S Suvin.

Table 2.5. Identification of compounds present in sample—GCF. Copyright Dr P S Suvin.

Sl. No.	GCF sample compounds	Inference
1	Bis(2-ethylhexyl) phthalate	H317: May cause an allergic skin reaction.
2	5-Hepten-2-one, 6-methyl	H226: Flammable liquid and vapour, H319: [Warning: serious eye damage/eye irritation].
3	Nonane, 2,6-dimethyl	Not enough information available.
4	D-Limonene	H226: Flammable liquid and vapour, H304: May be fatal if swallowed and enters airways, H315: Causes skin irritation/corrosion, H410 : Very toxic to aquatic life with long lasting effects.
5	2,6-Octadienal, 3,7-dimethyl-, (E)	H315: Causes skin irritation, H317: May cause an allergic skin reaction, H319: [Warning: serious eye damage/eye irritation].
6	Hexadecane	May be fatal if swallowed and enters airways.

Table 2.6. Identification of compounds present in a few samples—CCF. Copyright Dr P S Suvin.

Sl. No.	Elements in CCF	Hazardous nature
1	3-Carene	H226: Flammable liquid and vapour, H304: May be fatal if swallowed and enters airways, H315: Causes skin corrosion/ irritation, H317: May cause an allergic skin reaction [Warning: sensitization, skin], H412: Harmful to aquatic life with long lasting effects.
2	D-Limonene	H226: Flammable liquid and vapour, H304: May be fatal if swallowed and enters airways, H315: Causes skin irritation [Warning: skin corrosion/irritation], H317: May cause an allergic skin, H400: Very toxic to aquatic life, H410: Very toxic to aquatic life with long lasting effects

3	Isosorbide	Human exposure: Main risks and target organs: Vasodilatation and hypotension (with their accompanying complications) are the main risks with overdose of isosorbide dinitrate. Heart and blood vessels are the target organs. Methemoglobinemia can occur.
4	Longifolene	H304: May be fatal if swallowed and enters airways—danger: aspiration hazard, H317: May cause an allergic skin reaction, H400: Very toxic to aquatic life, H410: Very toxic to aquatic life with long lasting effects.
5	Hexadecane	May be fatal if swallowed and enters airways.
6	1-Octanol, 3,7-dimethyl	H315: Causes skin irritation [Warning: skin corrosion/irritation], H319: Causes serious eye damage/eye irritation, H411: Toxic to aquatic life with long lasting effects.
7	Ethanol, 2-(2-butoxyethoxy)	H227: Combustible liquid [Warning: flammable liquids], H319: Causes serious eye damage/eye irritation, H336: May cause drowsiness or dizziness [Warning: specific target organ toxicity], H372: Causes damage to organs through prolonged or repeated exposure [Danger: specific target organ toxicity, repeated exposure].
8	1-Octanol, 2-butyl	H400: Very toxic to aquatic life, H411: Toxic to aquatic life with long lasting effects.
9	Naphthalene, 1,6,7-trimethyl	H302: Harmful if swallowed [Warning: acute toxicity, oral], H400: Very toxic to aquatic life, H410: Very toxic to aquatic life with long lasting effects [Warning: hazardous to the aquatic environment, long-term].
10	Dodecane	H413: May cause long lasting harmful effects to aquatic life, may be fatal if swallowed and enters airways, H226: Flammable liquid and vapor.
11	Tridecane	Repeated or prolonged skin contact may irritate or redden skin, progressing to dermatitis. Exposure to high concentrations of vapor may result in headache. Tridecane may be harmful by inhalation, ingestion, or skin absorption during industrial use. When aspirated into the lungs, tridecane is an asphyxiant. It can cause death and chemical pneumonitis.
12	Tetradecane	H304: May be fatal if swallowed and enters airways. Acute/chronic hazards: Explosion hazard: Moderate, in the form of vapor when exposed.
13	Phenol, 2,4 bis(1,1-dimethylethyl)	Very toxic to aquatic organisms. Special care should be taken to assure that 2,4 DTBP or mixtures containing 2,4 do not enter the aquatic environment.
14	3-Eicosene, (E)-	Use personal protective equipment. Eyeshields, gloves, respirator filter.
15	Trichloroacetic acid, hexadecyl ester	Hazardous to the aquatic environment. acute hazard, chronic hazard, skin corrosion, serious eye damage.

Here follow certain findings from the GC–MS results. It was observed that a number of chemicals found to be toxic are present in both the GCF and the CCFs but GCF showing extremely safe levels of toxicity in the OECD-203 Acute Fish Toxicity Test and the Fish Embryo Test, so we can see that the amount or dosage of these harmful chemicals in the GCF are negligible to that present in the CCF.

2.3.3 Biodegradability and corrosion study of cutting fluids

The biodegradability and BOD-to-COD ratio were assessed using biochemical oxygen demand (BOD) and chemical oxygen demand (COD) tests. BOD is a measurement of the amount of dissolved oxygen required by aerobic living organisms to break down organic material contained in a water sample at a specific temperature over a specific time period. The amount of dissolved oxygen absorbed by aerobic bacteria over a five day period at 20 °C is referred to as BOD. The entire amount of organic materials in a waste stream is referred to this BOD. COD is the measure of oxygen required to chemically breakdown organic compounds in water by digesting with an acid. It is usually done by mixing appropriate chemicals into a water sample at a certain temperature for a set amount of time. The COD test is a widely used indirect method for determining how many organic chemicals are present in water. COD is a helpful measure of water quality since it determines the amount of organic contaminants contained in surface water (e.g., lakes and rivers) or wastewater. It is measured in milligrammes per litre (mg L^{-1}), which means how much oxygen is absorbed per litre of solution. The principle behind the test was that the organic matter in the sample gets oxidized completely by potassium dichromate ($K_2Cr_2O_7$) in the presence of sulfuric acid (H_2SO_4), silver sulfate ($AgSO_4$), mercuric sulfate ($HgSO_4$), to produce CO_2 and H_2O. The sample is titrated against ferrous ammonium sulfate using ferroin as indicator. The amount of dichromate consumed by the sample is the amount of oxygen required to oxidize the organic matter. The amount of oxygen consumed is directly proportional to the COD value.

The chemical and biological nature of wastewater is determined. Organic matter includes contaminants such as the GCF generated in our lab. Organic matter, often known as organic compounds, is a broad category of chemicals made up of carbon and one or more of the following elements: hydrogen, nitrogen, or oxygen, and originated from once-living matter. The oxygen demand of an effluent or wastewater discharge predicts its impact on the receiving water. The BOD and the COD are the two most prevalent parameters used to determine the composition of wastewater (COD). BOD is a measurement of how much dissolved oxygen aerobic bacteria eat in 5 days at 20 °C. It is a broad measure of the organic matter's strength in a waste stream. COD is determined chemically through acid digestion. Under some conditions, there is a clear correlation between COD and BOD, and by determining the COD, information about the BOD of the wastewater can be acquired, however it is very waste dependent. The biodegradability of GCFs, which are essentially made up of biodegradable materials, was examined alongside several CCF samples. COD

Table 2.7. BOD/COD for cutting fluids. Copyright Dr P S Suvin [9].

Sample	BOD (mg L^{-1})	COD (mg L^{-1})	BOD/COD
GCF	42	40	1.05
CCF1	78	68	1.14
CCF2	48	176	0.27
CCF3	98	224	0.4
CCF4	70	216	0.32
CCF5	54	360	0.15

analysis, which can be performed instead of the BOD test, predicts the quantity of organic matter in wastewater in three to four hours rather than the five days required by the BOD test. COD values are usually greater than BOD values, and the ratio between them varies based on the wastewater's characteristics. The experiment was carried out according to Indian standards (IS 3025: (Part 58)—Reaffirmed 2006 and IS: 3025 (Part 44)—Reaffirmed 2003) and under certain conditions. The values of BOD and COD of fresh and used samples were determined using titration-based methods.

Cutting fluid BOD and COD levels are shown in table 2.7.

The BOD to COD ratio gives the level of biodegradability of a sample:

- BOD/COD is > 0.6 = biodegradable.
- BOD/COD is between 0.3 and 0.6 = slow action, biodegradable.
- BOD/COD is < 0.3 = not biodegradable.

The biodegradability property was investigated by comparing the BOD-to-COD ratio. GCF was found to be biodegradable, while CCF was found to be non-biodegradable. Our GCF has a BOD/COD ratio of 1.05, indicating that it is highly biodegradable, whereas the BOD/COD ratios of the five commercial fluids examined were determined to be non-biodegradable. GCF was shown to be biodegradable and non-toxic to the environment, in addition to being a long-term product created from renewable resources.

2.3.4 Corrosion aspect of MWF (corrosion tests)

When freshly cut surfaces are formed during metal machining, the anti-corrosive quality of cutting fluid is an important property to test. Cutting fluid samples were tested for anti-corrosion properties using (1) ASTM D4627 criteria and (2) a weight loss method developed in-house. During the experiment, 5 mL of cutting fluid was poured onto 4 g of cast iron chips in a petri dish, which were placed on filter paper. The cutting fluid was drained after 24 h of incubation and left to stand for 72 h. The degree of rusting stain on the filter paper was used to determine the anti-corrosion characteristics of cutting fluids [19]. Corrosion of cast iron chips resulted in proportional weight loss, therefore, the weight loss of the cast iron chips after cleaning them with acetone for 20 min was determined at 24 and 72 h. The weight of the cast iron chips was measured after drying, and weight loss due to corrosion was investigated. According to ASTM D4627, the rate of corrosion was evaluated on a scale of one to ten, with one indicating non-corrosive properties, five indicating medium, and ten indicating the most corrosive nature. When compared to grade 4 of most CCF samples examined, the tests show that GCF with grade 3 has greater anti-corrosive properties. Figures 2.12 and 2.13 show the results of the observations. After a 24-h incubation period, the new GCF (20 mg) shows greater corrosion resistant qualities when compared to CCF (25 mg), according to the weight

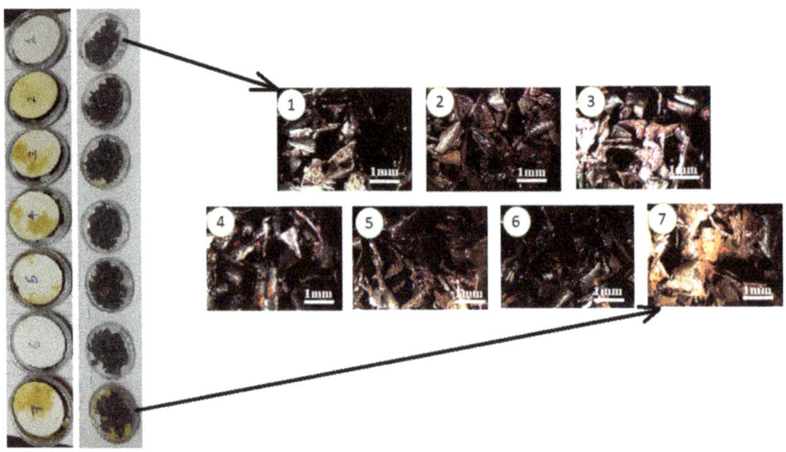

Figure 2.12. After 24 h of testing, the chips and discoloration on the filter paper were seen. CCF is ranked 1–5, GCF is ranked 6–7, and control is ranked 8–10 (deionized water). GCF and CCF1 showed no corrosion in the filter paper, whereas all other CCFs and the control showed corrosion stains. Copyright 2021 SAGE Publications [15].

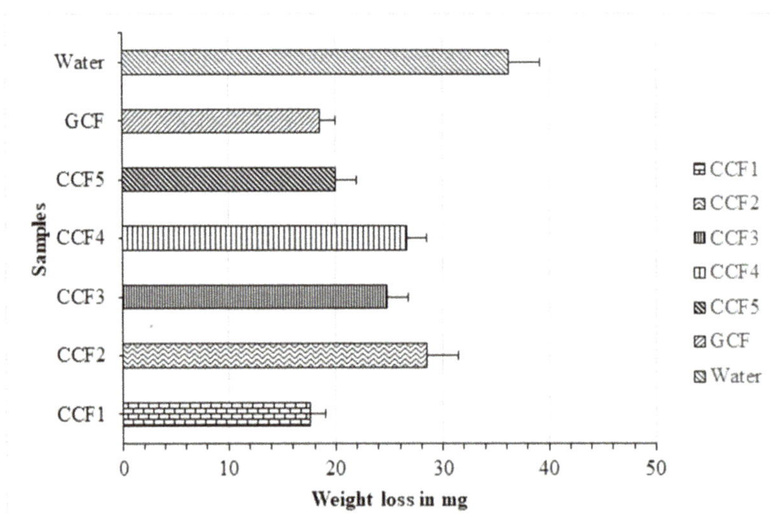

Figure 2.13. Weight loss results of cutting fluids tested 1–5 are the different CCFs, 6-GCF and 7—control (deionized water). Copyright 2021 SAGE Publications [15].

loss study. The results showed that the GCF has a corrosion rate that is comparable to, if not better than, the CCF.

Generally, the efficiency of inhibitors on the steel surface is controlled by several factors such as: (a) number of the adsorption sites and their charge density, (b) nature of interaction between inhibitors and the metal surface, etc. Figure 2.14

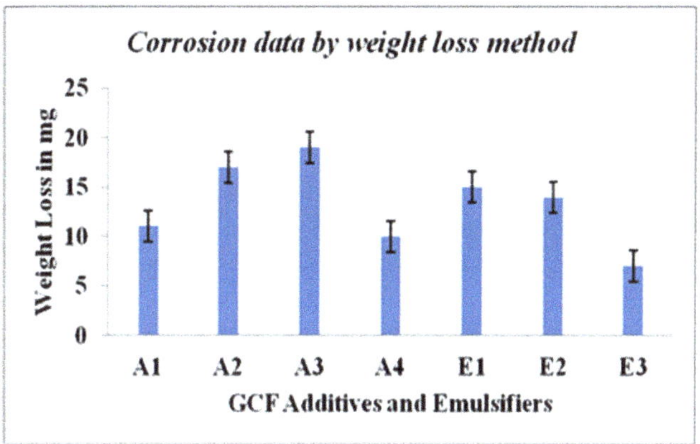

Figure 2.14. Weight loss results of ingredients used in cutting fluids. A1, A2, A3, A4, E1, E2, and E3. Copyright Dr P S Suvin.

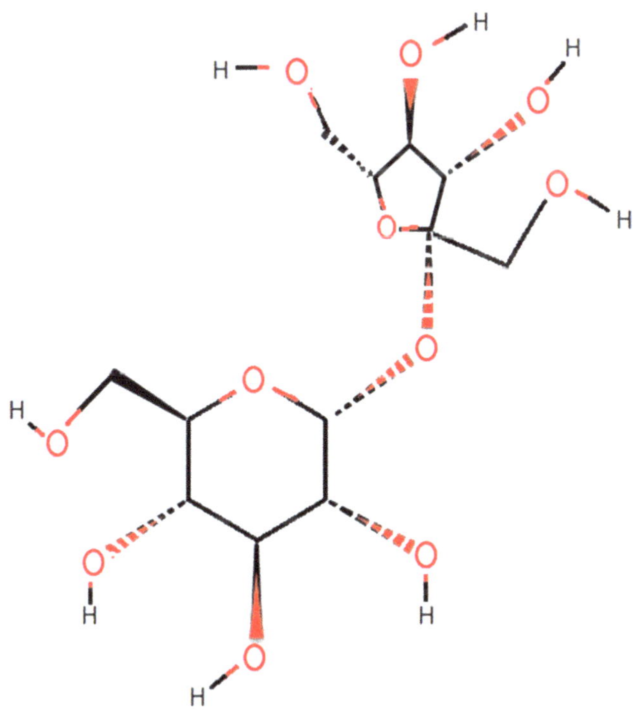

Figure 2.15. Sucrose compound structure. Copyright Dr P S Suvin.

Table 2.8. E2 composition. Copyright Dr P S Suvin.

Composition of E2	
Acid name	Composition range
Linoleic acid	6%–16%
Oleic acid	25%–54%
Hexadecanoic acid	16%–33%
Octadecanoic acid	9%–24%

shows the weight loss test results for different ingredients used in GCF preparation, tested for studying the corrosion inhibition property. Additives A1 and A4 and emulsifier E3 showed better corrosion inhibition properties compared to other ingredients.

Reasons for the corrosion inhibition efficiency of GCF are as follows:

(1) E3 is corrosion preventative. The presence of a terminal amine with a larger electron density that binds more strongly to the metal surface is responsible for E1's high inhibitory efficiency [20].

(2) Reason for less inhibition in A4 is because it contains 50%–60% sucrose.

The lone pair of electrons found in oxygen may aid in sucrose adsorption on the surface. Figure 2.15 illustrates this.

(3) Reason for less inhibition of A2 is because of fatty acids which adsorb on the surface and act as a barrier between solution and substrate. Table 2.8 shows the composition of A2.

(4) The natural oils chosen have several fatty acids that could be adsorbed on the surface giving the protection required.

2.3.5 Bacterial growth study

Being oil-in-water emulsions, MWFs are prone to growth of bacteria and fungi and hence microbial tests were also performed to study their effect on exposure to workers. Therefore, bacteriological analysis of the GCF was carried out using different culture and plating techniques, and it was found that combination of the green additives (A1, A2, and A3) inhibited the growth of microorganisms synergistically and yielded results similar to those of the CCFs.

Microbiology of cutting fluid:

- microorganisms can foul cutting fluids;
- they interfere with the desired properties of the oil;
- objectionable odors are formed;
- they may contribute to dermatitis, skin infections and other diseases;
- metallic surfaces may be corroded.

Here, used cutting fluid emulsion is taken from all the cutting fluids and known bacteria are inoculated into the sample to see the antimicrobial nature of the cutting fluids to the induced bacteria. Here fresh emulsion is prepared of all the cutting fluids and known bacteria are inoculated into the sample to see the antimicrobial nature of the cutting fluids to the induced bacteria. The procedure for the test is shown in figure 2.16.

Species 1: *Proteus* sp.

Species 2: *Pseudomonas* sp.

Species 3: *Shigella* sp.

Sample plates arrangement before incubation, plates after sterility check and plates after growth of bacteria are shown in figures 2.17, 2.18, and 2.19, respectively. Tables 2.9, 2.10 and 2.11 show the CFU mL^{-1} of each species.

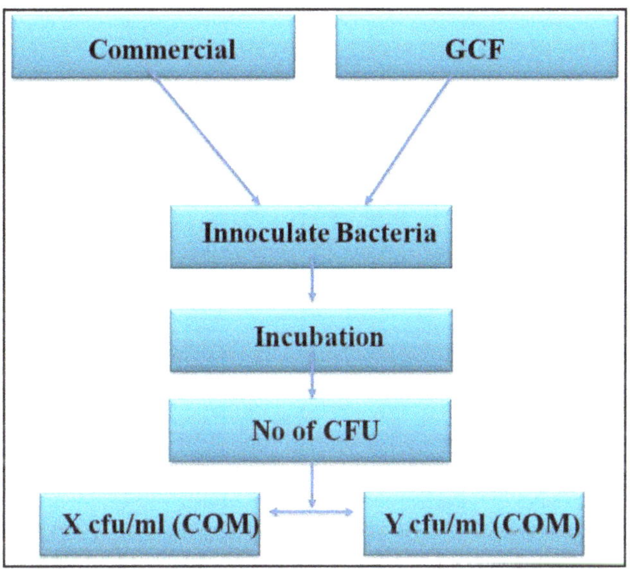

Figure 2.16. Schematic diagram for the procedure to be followed. Copyright Dr P S Suvin.

Figure 2.17. Plates before incubation. Copyright Dr P S Suvin.

Figure 2.18. Plates after sterility check. Copyright Dr P S Suvin.

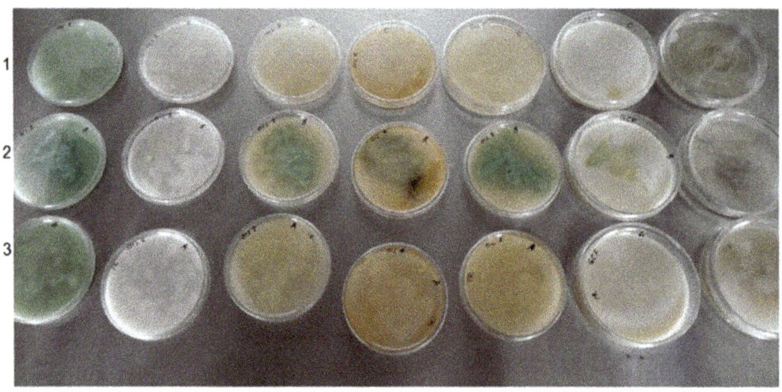

Figure 2.19. Plates after growth of bacteria. Copyright Dr P S Suvin.

Table 2.9. Observations from plate incubated with species 1. Copyright Dr P S Suvin.

Plate number	No. of colonies	CFU mL^{-1}
CCF1	TNC	–
CCF2	TNC	–
CCF3	Nil	–
CCF4	100	$3 \times 3 \times 10^3$
CCF5	198	$3 \times 6.6 \times 10^3$
GCF	32	$3 \times 1.06 \times 10^3$
Control	50	$3 \times 1.67 \times 10^3$

SPECIES 1: *Proteus* sp.

Contamination: Nil

The growths of bacterial colonies on the petri plates were counted. It was observed that CCF3 had no growth and the growth of the bacteria *Proteus* sp. was

Table 2.10. Observations from plate incubated with species 2. Copyright Dr P S Suvin.

Plate number	No. of colonies	CFU mL^{-1}
CCF1	118	$3 \times 3.93 \times 10^3$
CCF2	Nil	–
CCF3	TNC	–
CCF4	142	$3 \times 4.7 \times 10^3$
CCF5	TNC	–
GCF	61	$3 \times 2.03 \times 10^3$
Control	90	$3 \times 3 \times 10^3$

Table 2.11. Observations from plate incubated with species 3. Copyright Dr P S Suvin.

Plate number	No. of colonies	CFU mL^{-1}
CCF1	TNC	–
CCF2	Nil	–
CCF3	293	$3 \times 9.76 \times 10^3$
CCF4	TNC	–
CCF5	210	$3 \times 7 \times 10^3$
GCF	90	$3 \times 3 \times 10^3$
Control	110	$3 \times 3.6 \times 10^3$

the least in GCF with count of 32 colonies per plate. This implies that the GCF is giving comparable antimicrobial property to the commercially available cutting fluids. TNC represents too numerous to count.

SPECIES 2: *Pseudomonas* sp.
Contamination: Nil

The growths of bacterial colonies on the petri plates were counted. It was observed that the growth of the bacteria *Pseudomonas* sp. was the least in GCF with count of 61 colonies per plate. This implies that the GCF is better for sustainable use as compared to the commercially available cutting fluids.

SPECIES 3: *Shigella* sp.
Contamination: Nil

The growths of bacterial colonies on the petri plates were counted. It was observed that the growth of the bacteria *Shigella* sp. was the least in GCF with count of 90 colonies per plate. This implies that the GCF is better for sustainable use as compared to CCFs. The GCF used in this study was developed after several candidate GCFs were synthesised, analysed and discarded on account of their low biodegradability rates and their inability to provide proper lubrication with cooling properties. They were also found to be more toxic than the GCF used in this study. The lubricant used in this study proliferated with less microbial growth compared to the other synthesised oils. This reduces its toxic effect on the health of living organisms and its effect on biomagnification. The tests carried out concluded that the synthesized GCF had comparable bacterial growth to the plates with commercially synthesized oil (CCF). When tested under GC–MS for the components (explained in section 2.3.2), it was found that the CCF did have high levels of toxic compounds and elements like phenol, limonene, naphthalene, longifolene, etc. This proves the last of our conclusions, i.e. the high toxicity of the commercial fluid inhibited the growth of the microbes.

2.3.6 Drilling and tool chip tribometer test

2.3.6.1 Drilling test

Drilling tests were performed to assess the machining performance of various cutting fluids by measuring cutting forces during these metal cutting operations. Machining performance on drilling was determined by measuring the axial force and feed torque. The cutting forces were measured using strain gauge type drill dynamometer mounted on a specially designed fixture. The controller will display the thrust force and torque measured [16]. This was performed using an HMT FN2 universal milling machine. A 200 mm × 65 mm × 40 mm AISI 1018 mild steel workpiece was used. The workpiece was immersed while the holes were bored. The force and torque were measured using a drill dynamometer from Magnum Instruments. Holes of 30 mm depth were drilled using a 10 mm diameter HSS drill bit (HSS drill bit of diameter 10 mm, flute length 85 mm, point angle 1200°) for various combinations of controllable parameters speed (355 rpm, 710 rpm, 1120 rpm), feed (4 mm min^{-1}, 10 mm min^{-1}, and 25 mm min^{-1}), with each experiment repeated four times [15].

The results of these tests revealed that the GCF's performance was comparable to that of the CCFs.

Various tests were carried out in which one parameter was changed while all other factors, such as cutting speed, cutting feed, and workpiece material, remained fixed. Cutting force and torque necessary or consumed for a certain set of trials were measured and analysed in these experiments.

For each set of studies, the cutting fluid was changed from mineral to vegetable based to see whether fluid performed better in terms of reduced force and torque. The test results are not absolute values, but rather compare the performance of the test fluid to that of a reference fluid. It was discovered that by using cutting fluid, the force and torque values were reduced. Because cutting fluid acts as a lubricant at the contact between the tool and the workpiece, it reduces friction. The thrust force value reduced when the spindle speed was raised, but the thrust force increased while the feed rate was increased, as indicated in the graphs in figures 2.20 and 2.21. Because material removal is high, increasing the feed rate raised the surface roughness value. The surface roughness value was lowered as the spindle speed was increased. According to the findings of drilling studies, the newly formed GCF is equivalent to CCF in machining processes and could be a viable replacement in the cutting fluid industry.

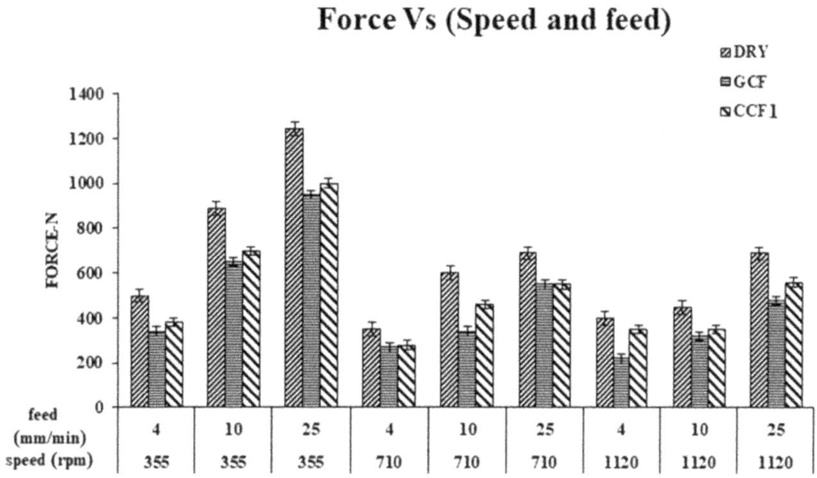

Figure 2.20. For dry drilling and drilling with CCF and GCF, axial/thrust force varies with speed and depth of cut. Copyright 2021 SAGE Publications [15].

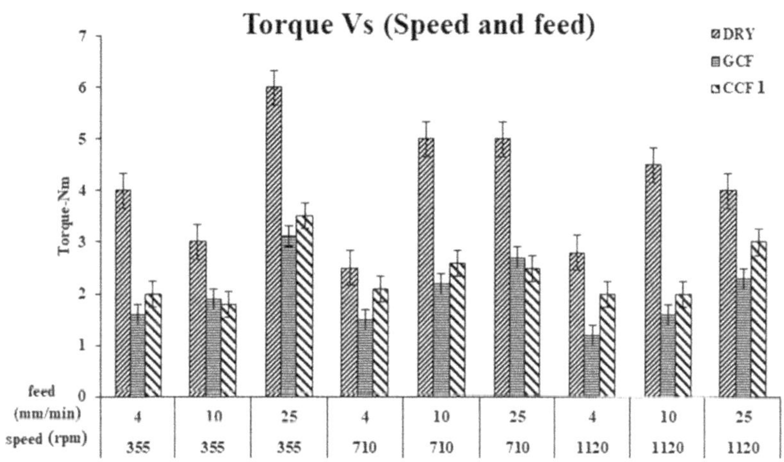

Figure 2.21. For dry drilling and drilling with CCF and GCF, torque varies with speed and feed. Copyright 2021 SAGE Publications [15].

GCF drilling experiments yielded comparable findings. GCF even outperformed CCF for a few factors. As a result, the investigations demonstrated that the properties of prepared GCF were comparable to those of CCF [15].

2.3.6.2 Tool chip tribometer

The experimental inquiry made use of a one-of-a-kind tribometer that can perform friction testing on freshly sliced surfaces. The experimental setup is comparable to that of a pin-on-disc tribometer. Experiments are carried out in this experimental facility by performing cutting operations inside a lubricating pool. The disc rotates around a vertical axis and engages with a tool to cut a circular cut track on the flat surface of the disc that is nearly 4 mm wide. As the cutting occurs inside the pool, the lubricant film on the cut track will be due to the action of the fluid on a nascent surface, similar to the action of cutting fluids on newly generated surfaces during machining and friction force is measured *in situ* by the ball mounted on a pin fixed to a friction-force measuring arm positioned such that it engages against the rotating disc along the cut track generated by the tool. Figure 2.22 shows the setup. Experiments were performed at a fixed load of 1 kg and various speeds of 30, 50 and 70 rpm. The task is determining the coefficient of friction between a pin and a nascent surface formed by a single point tool cutting metal. To create the nascent surfaces, the disc was designed to connect with a tool that was placed in a fluid-filled tank. The time lag between the moment of cutting and the point of measuring is determined by the disc's rpm. The disc surfaces stayed 2 mm below the liquid level in the tank at all times. A cut was made that was 3 mm deep and 5 mm wide. Friction measurements on the sliced surface have been taken *in situ*. There were two types of materials used (aluminum and MS). Along with GCF, commercial cutting fluid was used. Formulations were examined using a series of tests in which one parameter was varied while all other parameters, such as cutting speed and workpiece material, remained fixed. The HSS tool was used to conduct experiments at 30, 50, and 70 rpm. These investigations tested and studied the frictional force discovered for a certain set of studies. For each set of studies, the cutting fluid was changed from mineral to vegetable based to see which fluid performed better in terms of reduced frictional force and good surface quality. The test findings are not absolute values, but rather compare the efficiency of the test fluid to that of a reference fluid or to that of other fluids. The test entails continually cutting an uncut surface, continuously measuring the coefficient of friction (COF) after tool engagement, continuously disengaging the tool, and continuously measuring the COF after tool withdrawal [21]. A smooth transition from a condition of high COF to a state of low COF is noticed when cutting is halted (by withdrawing the tool) without interrupting the pin sliding against the surface [22].

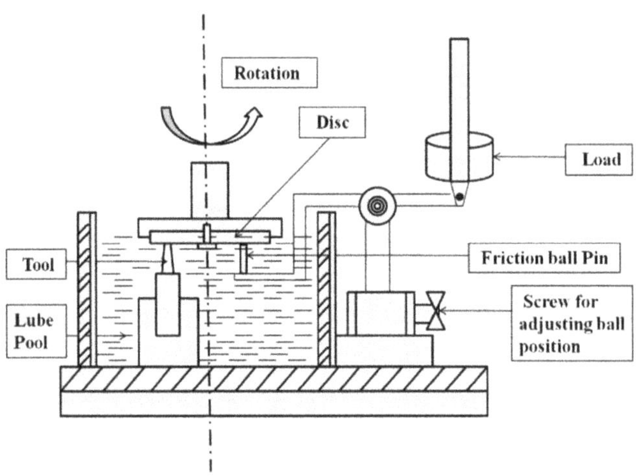

Figure 2.22. Tool chip tribometer schematic. Copyright 2021 SAGE Publications [15].

In characterization and analysis study—the ball scar diameter, the frictional force and surface roughness have been compared. Thus the performance was evaluated with respect to these properties. Figure 2.23 gives the COF variation with time for 30 rpm for aluminium as workpiece. Frictional response for both GCF and COM was recorded and it was observed that the COF of nascent surface is showing highest level of friction as the newly generated or nascent surface will be more sensitive and the surface is not oxidised. GCF and CCF are found comparable with results obtained. Figure 2.24 gives the COF variation with time for 30, 50 and 70 rpm, for

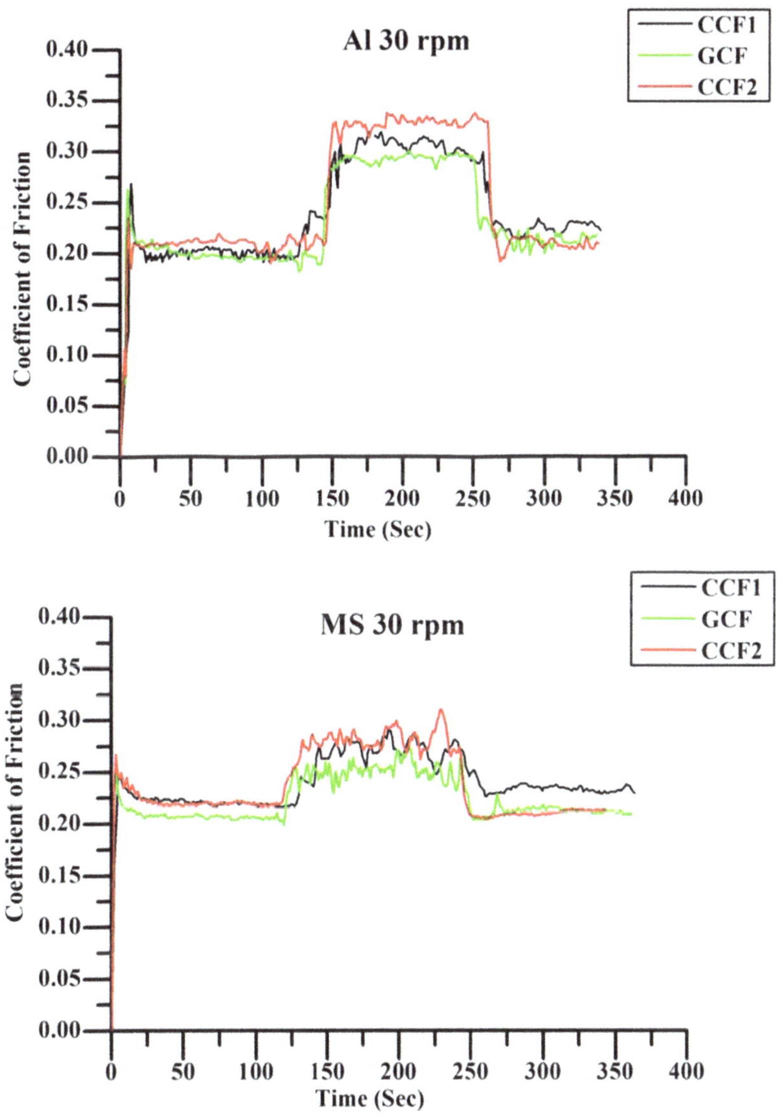

Figure 2.23. COF variation with time compared using GCF, CCF1, and CCF2 as the cutting fluids for MS and Al disc with 30 rpm speed. Copyright 2018 by Japanese Society of Tribologists [23].

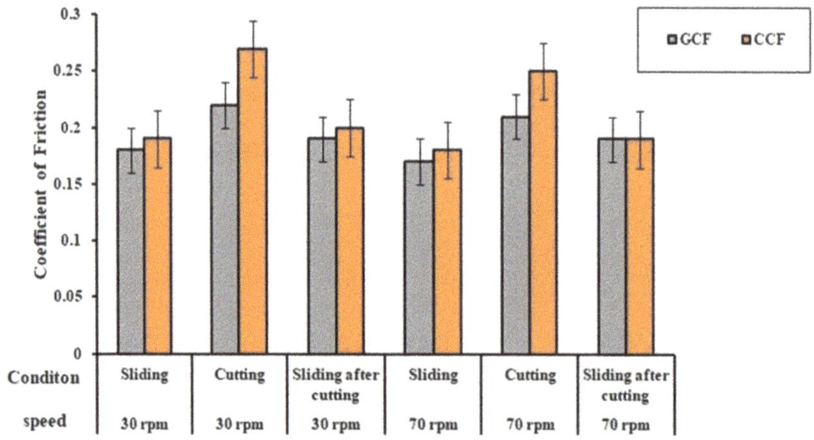

Figure 2.24. Consolidated MS results showing coefficient of friction values of GCF giving comparable results with CCF. Copyright 2018 by Japanese Society of Tribologists [23].

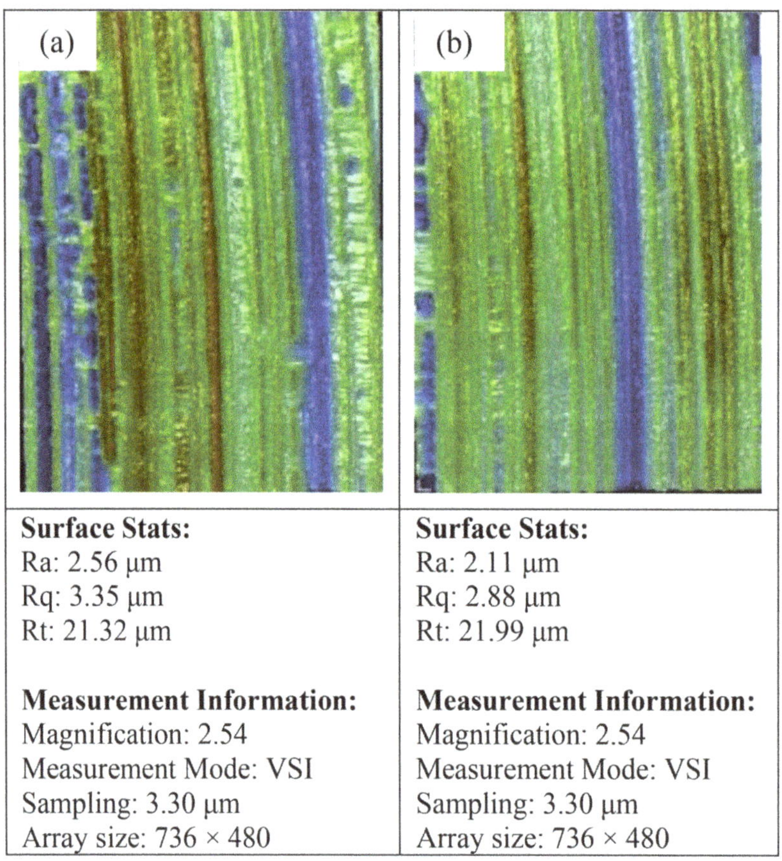

Figure 2.25. Optical profiler image of MS sample machined at 70 rpm using CCF and GCF. Copyright 2018 by Japanese Society of Tribologists [23].

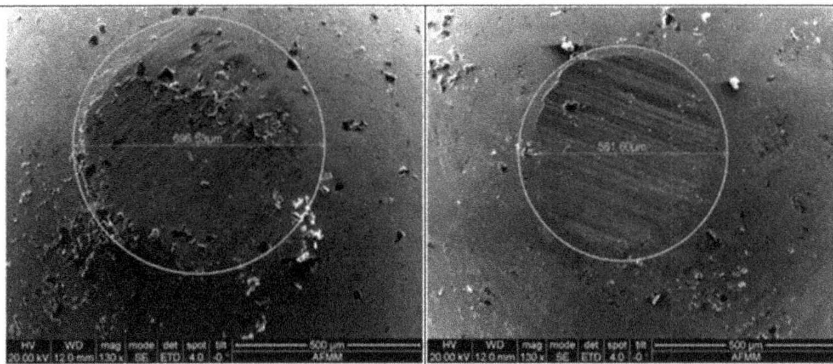

Figure 2.26. SEM image of wear mark on HSS ball after sliding with mild steel disc—70 rpm for CCF and GCF. Copyright 2018 by Japanese Society of Tribologists [23].

mild steel as workpiece. COF values for MS disc were found less compared to Al disc as there is less friction between tool and workpiece interaction in the case of MS disc. Increase in sliding speed may sometimes result in starvation as there won't be enough time for the formation of boundary film. In the case of MS disc also recorded COF values are comparable between GCF and COM.

All three surfaces have different COFs, with the nascent surface having the highest level of friction. The oxide layer on the ball fixed to the pin, as well as that generated during the cleaning of the disc after it has been cut, are both effective at combining with the oil during the rubbing of the 'oxidised' surface to yield organic compounds that form an antifriction film at the interface in the 'rubbing' test. From the analysis of cutting fluid performance in tool chip tribometer operation, frictional force and roughness measured was found to be comparable with CCF. The part (dimensional) accuracy was in some cases better than mineral-based cutting fluid. For example, a roughness optical profilometer image for a cutting in ms workpiece with 70 rpm for CCF and GCF is shown in figure 2.25 and corresponding wear scar in figure 2.26. By using cutting fluid we can reduce the tool–work interface temperature, reduce the friction between workpiece and the tool and in turn improve the surface finish. Low COF results from change in shear rate by increase in sliding speed. The influence of tribofilm formed by oil between tool and work formed by each cutting oil, could have brought the difference in results.

2.4 Conclusion

Green MWF made with coconut oil and a variety of non-toxic emulsifiers and natural additives satisfies many of the characteristics of conventional formulations and even outperforms them in terms of environmental impact. DLS and pH stability test indicate the formulated emulsion is stable. GCF gave comparable machining performance and good anti-corrosion properties. GCF promotes healthy work environments and prevents workplace illnesses, compared to the side effects caused due to petroleum-based cutting fluids. To make perfect choice of fluids, it is important to consider not only the fluids' performance characteristics, but other factors, such as waste treatment, disposal cost, fluid life, resistance to microbial attack etc. Nevertheless, this formulation is a good basis for further development and utilization of bio-lubricants for various applications.

References

[1] Brinksmeier E, Meyer D, Huesmann-Cordes A G and Herrmann C 2015 Metalworking fluids—mechanisms and performance *CIRP Ann.* **64** 605–28

[2] Marksberry P W and Jawahir I S 2008 A comprehensive tool-wear/tool-life performance model in the evaluation of NDM (near dry machining) for sustainable manufacturing *Int. J. Mach. Tools Manuf* **48** 878–86

[3] Ojolo S J and Ohunakin O S 2011 Study of rake face action on cutting using palm-kernel oil as lubricant *J. Emerg. Trends Eng. Appl. Sci.* **2** 30–5

[4] Xavior M A and Adithan M 2009 Determining the influence of cutting fluids on tool wear and surface roughness during turning of AISI 304 austenitic stainless steel *J. Mater. Process. Technol.* **20** 900–9

[5] Najiha M S, Rahman M M and Yusoff A R 2016 Environmental impacts and hazards associated with metal working fluids and recent advances in the sustainable systems: a review *Renew. Sustain. Energy Rev.* **60** 1008–31

[6] Greeley M and Rajagopalan N 2004 Impact of environmental contaminants on machining properties of metalworking fluids *Tribol. Int.* **37** 327–32

[7] Fox N J and Stachowiak G W 2007 Study of hazards related to cutting fluids and their remedies *Tribol. Int.* **40** 1035–46

[8] Lawal S A, Choudhury I A and Nukman Y 2012 Application of vegetable oil-based metalworking fluids in machining ferrous metals—a review *Int. J. Mach. Tools Manuf* **52** 1–12

[9] Shihab S K, Khan Z A, Aasmohammad and Siddiquee A N 2013 Application of response surface methodology for determining cutting forces in hard turning using castrol coolant *Int. J. Adv. Mater. Manuf. Character.* **3** 27–36

[10] Singh Y, Singla A, Singh A K and Upadhyay A K 2018 Tribological characterization of *Pongamia pinnata* oil blended bio-lubricant *Biofuels* **9** 523–30

[11] Lawal S A, Choudhury I A and Nukman Y 2013 Developments in the formulation and application of vegetable oil-based metalworking fluids in turning process *Int. J. Adv. Manuf. Technol.* **67** 1765–76

[12] Fox N J and Stachowiak G W 2007 Vegetable oil based lubricants—a review of oxidation *Tribol. Int.* **40** 1035–46

[13] Jayal A S, Badurdeen F, Dillon O W Jr. and Jawahir I S 2010 Sustainable manufacturing: modelling and optimization challenges are the product, process and system levels *CIRP J. Manuf. Sci. Technol.* **2** 144–52

[14] Menezes P L, Nosonovsky M, Ingole S P, Kailas S V and Lovell M R 2013 *Tribology for Scientists and Engineers* (Cham: Springer)

[15] Suvin P S, Poorva G, Horng J H and Kailas S V 2020 Evaluation of a comprehensive non-toxic, biodegradable and sustainable cutting fluid developed from coconut oil *Proc. Inst. Mech. Eng. J: J. Eng. Tribol.* **235** 1842–50

[16] Somashekaraiah R, Suvin P S, Gnanadhas D P, Kailas S V and Chakravortty D 2016 Eco-friendly, non-toxic cutting fluid for sustainable manufacturing and machining processes *Tribol. Online* **11** 556–67

[17] O E C D *Test No. 203: Fish, Acute Toxicity Test* OECD Publishing

[18] Dayrit F M, Buenafe O E M, Chainani E T, de Vera I M S, Dimzon I K D, Gonzales E G and Santos J E R 2007 Standards for essential composition and quality factors of commercial virgin coconut oil and its differentiation from RBD coconut oil and copra oil *Philipp. J. Sci.* **136** 119–29

[19] ASTM D4627 2012 *Test Method for Iron Chip Corrosion for Water-Dilutable Metalworking Fluids* (West Conshohocken, PA: ASTM International)

[20] Watanabe S, Nakagawa H, Ohmori Y, Fujita T, Sakamoto M and Haga T 1996 Characteristic properties of cutting fluid additives derived from half esters of dimer acids *J. Am. Oil Chem. Soc.* **73** 807–09

[21] Anirudhan P, Kailas S V and Biswas S K 2011 Lubricity of an oil in water emulsion in metal cutting: the effect of hydrophilic/lypophilic balance of emulsifiers *Colloids Surf.* A **384** 323–30

[22] Anirudhan P, Kailas S V and Biswas S K 2010 Experimental estimation of friction force in lubricated cutting of steel *Wear* **269** 557–64

[23] Suvin P S and Kailas S V 2018 Study and comparison of lubricity of green and commercial cutting fluid using tool-chip tribometer *Tribol. Online* **13** 340–50

Chapter 3

Tribology of multifunctional bio-based lubricants

Sougata Roy

3.1 Introduction

One of the growing concerns regarding mineral oil-based lubricants is their detrimental effect on the environment: an oil and gas leak incident in 1969 contaminated a Santa Barbara, California beach with thirty miles of thick sludge. This incident was horrific and spurred an investigation into the environmental effects with a necessity of biodegradable oils. Base lubricant stocks can be categorized into three types: (a) petroleum or mineral-based, (b) synthetic, and (c) bio-based. 'Bio-based' lubricants are formulated using renewable and biodegradable base stocks. Bio-based oils do not necessarily need to be unaltered vegetable oil; however, the base material should be renewable. Tribology is the science of friction, wear, and lubrication characteristics existing between interacting surfaces that are in relative motion to each other. Understanding the tribological behavior of bio-based lubricants is critical to expanding their applicability in various sectors since lubricants are used to reduce friction between moving components. Friction must be as low as possible to avoid increased energy loss and reduced fuel economy, as in the case of cylinder liners and pistons used in internal combustion engines. High friction will increase the temperature at the contact surface, resulting in the seizure and complete failure of the component.

Wear is defined as the surface damage in one or more contacting surfaces due to relative motion. The gradual removal of materials due to surface abrasion generally occurs through asperity contacts. A range of wear modes, such as abrasion, adhesion, fatigue, fretting, corrosion, and oxidation, are possible depending on contact conditions and environmental factors. One of the primary purposes of applying a lubricant is to reduce friction and wear in contacting surfaces. However, if the lubricant reacts adversely with contacting surfaces, it can result in increased friction or increased wear or both. Preliminary tests should be carried out to understand antagonistic or synergistic behavior of different additive packages mixed

doi:10.1088/978-0-7503-3435-8ch3

in lubricant. In automotive engines, the lubricants generally help to transfer the generated heat due to reciprocating motion between piston rings and cylinder liners. Suitable additives in lubricants can form tribofilm on the contacting surfaces to protect against different wear modes. Depending on contact surface conditions the lubricant film thickness can vary resulting in boundary, mixed, elasto-hydrodynamic and hydrodynamic lubrication.

Energy demand has increased due to successful industrialization, development, and modernization within various sectors. Petroleum-based fuels and lubricants currently supply most global energy requirements, which are expected to increase by as much as 33.5% by 2030 [1]. For every 30–40 million tons of lubricants used, 55% of the total amount would be back to the environment [2], 95% of which are petroleum-based and harmful to the environment [3]. Fossil fuel emissions may increase global temperatures by 1.4 °C to 5.8 °C from 1990 to 2100, as reported by the Intergovernmental Panel on Climate Change (IPCC) [4]. Petroleum-based lubricants are a toxic and non-renewable resource that may not be obtainable in the near future as the supply is exhausted, a significant concern for adequate energy production; therefore, the search for potential petroleum-based lubricant alternatives has been initiated to mitigate these future concerns [5, 6].

Commercial lubricants used in the automotive sector are still mainly based on petroleum-based lubricants. 'Bio-based' lubricants, or 'bio-lubricants,' are derived from bio-based raw materials, such as animal fat, vegetable oil, or other environmentally friendly hydrocarbons [7]. These biodegradable and renewable lubricants offer good lubricity, high flash points, high viscosity indices, and enhanced shear resistance compared to petroleum-based products [8]. There are a few disadvantages to bio-oils, which hinder commercialization: poor oxidation stability, poor performance at low temperatures, and low viscosity index [9, 10]. We will discuss the physiochemical properties of bio-based lubricants in this chapter. Bio-oils are widely available, inexpensive, and highly biodegradable compared to petroleum-based lubricants [11]. Key tribological testing types for various bio-based lubricants will be reviewed, and the tribological characteristics of several bio-based lubricants will be analyzed.

3.2 Sources of bio-based lubricants

Bio-lubricants, classified as natural or synthetic oils, are derived from a wide range of bio-based feedstocks, depending on the feedstock type. Vegetable oils or animal fats, such as high-oleic sunflower and canola oils, are used to create natural oils that have high oxidation stability. Natural oils are used as a starting agent to develop synthetic oils with enhanced lubrication properties. The physiochemical properties of some popular vegetable oil-based lubricants are listed in table 3.1 [12, 13].

Some modified vegetable oil-based lubricants exhibit enhanced physiochemical properties (table 3.2).

Oils extracted from various vegetables, seeds, fruits, and animal fats are processed through several extraction and distillation steps [14, 15]. Several research groups have analyzed the physiochemical properties of saturated and unsaturated fatty

Table 3.1. Physiochemical properties of some vegetable oils. Information collected from [12, 13].

Lubricant	Viscosity at 40 °C (cSt)	Viscosity at 100 °C (cSt)	Viscosity index	Flashpoint (°C)	Pour point (°C)
Castor	220.6	19.72	220	250	−27
Soybean	28.86	7.55	246	325	−9
Olive	39.62	8.24	190	318	−3
Sunflower	40.05	8.65	206	252	−12
Coconut	24.8	5.5	169	325	21
Jatropha	40	5.5	170	273	−3
Rice bran	40.6	8.7	201	318	−13
Sesame	27.33	6.3	193	316	−5
Corn	32.41	8.06	238	324	−14
Rapeseed	45.6	10.07	180	252	−12

Table 3.2. Physiochemical properties of some modified vegetable oils. Information collected from [12, 13].

Lubricant	Viscosity at 40 °C (cSt)	Viscosity at 100 °C (cSt)	Viscosity index	Flashpoint (°C)	Pour point (°C)
Palm TMP ester	40.03	9.15	221	228	−36
Jatropha TMP ester	60.83	10.21	156	290	−35
Epoxidized soybean	162.8	19.3	136	300	−3
Epoxidized Jatropha	146.5	18.2	139	296	0

acids for use as green solvents. Four key chemical modification routes are used to modify vegetable oils. (1) Esterification or transesterification, where the transformation of an ester with enhanced thermal stability is conducted. This process improves oxidation stability but requires feedstocks with high-oleic acid content. (2) Hydrogenation, where the unsaturated and thermochemical cleavage of the ester is hydrogenated. This process reduces the degree of unsaturation and improves oxidation stability but requires a high reaction temperature. (3) Epoxidation, where a new epoxide functional group is introduced from double bond removal between two carbon atoms via an oxygen atom. This process improves lubricity and thermo-oxidative stability but increases the pour point value and decreases the viscosity index. (4) Estolide formation, where reactions between two identical or different acidic molecules take place. This process improves thermo-oxidative stability, but the production cost is higher compared to other processes.

3.3 Classification of bio-based lubricants

There are broadly three ways to categorize bio-based lubricants: (a) pure or chemically modified vegetable oil-based lubricants; (b) organic or inorganic additives mixed with pure vegetable oil-based lubricants; and (c) bio-based ionic liquids. Krzan (2010) indicated that in boundary lubrication regimes, pure vegetable oil-based lubricants exhibited enhanced lubricity compared to petroleum-based lubricants [16]. The author conducted nine oilseed crop experiments. Triglyceride vegetable oil structure exhibited enhanced wear protection, up to 1 GPa with a substrate temperature of 50 °C. Fatty acid molecules within vegetable oils can chemically react with metal surfaces to form a low shear strength metallic soap layer, which can significantly reduce the friction coefficient between surfaces [17].

The addition of various additives can improve the physiochemical properties of lubricants by increasing resistance to oxidation, viscosity, corrosion, rust inhibition, and decreased pour temperature [18, 19]; however, additives containing metals such as phosphorus, sulfur, or chlorine can be harmful to the environment and increase the toxicity of lubricants [20, 21]. Polymers, ionic liquids, plant-derived compounds, and layered and particulate materials are used extensively as alternative pure vegetable oil additives to avoid harmful environmental effects.

Fatty acids are chemically modified in bio-based vegetable oils to enhance their thermal and oxidative stability [14, 22]. Oils containing long-chained fatty acids, C18 and higher, exhibit an enhanced tribological behavior by forming a dense fatty acid monolayer that reduces asperity contacts and protects metal surfaces [23]. Jagadeesh and Satish indicated that fatty acid addition may exhibit inadequate surface protection (2012) [24]. The authors observed that increased wear may be caused by the corrosive effect of peroxides and oxidation-induced free fatty acids. Plant-derived cystine Schiff base ester exhibits significant anti-corrosion, anti-wear, and anti-friction properties due to disulfide groups (–S–S–) that form protective tribofilms on metal part contacts [25]. Inorganic non-toxic oxide nanoparticles, such as ZnO, CuO, TiO_2, reduce asperity contacts by developing smooth and thin surface protective films during sliding operations [26, 27]. BN, graphene, MoS_2, and WS_2 inorganic additives are added to natural and modified vegetable oils to create anti-wear or extreme pressure protective behaviors [26]. These additives have the unique advantage of acting as solid lubricants that can protect surfaces at high pressures and temperatures [15, 28, 29]. The addition of solid lubricant nanoparticles to vegetable oils allows for continuous contact [30–32], while the additized solution presents enhanced tribological characteristics.

Ionic liquids (IL) are an emerging class of bio-lubricant additives that can be used for various applications. Room-temperature ionic liquids are molten salts consisting of organic anions with melting points lower than 100 °C. These anions are paired with asymmetric organic cations with melting points of up to 300 °C [32]. Ionic liquid-based research has evolved significantly since its inception. The first widely known IL was discovered by Paul Walden (1914) more than 100 years ago, when he created ethyl ammonium nitrate [$(C_2H_5)NH_3NO_3$ or EAN] with a melting point of 12 °C [33]. In the 1970s, ILs were used in batteries as electrolytes [34–36]. A report from 2001 discussed the use of ILs as a lubricant [37]. This report was followed by studies using ILs as a neat lubricant [38–41]; however, this use was economically constrained despite the substance's advantageous physiochemical properties [42]. Oil-miscible ILs were developed that presented anti-scuffing and anti-wear properties, another new direction for the liquid's research [43, 44]. ILs have several unique characteristics, such as low flammability, low volatility, negligible vapor pressure, and high thermochemical stability. Bio-based ILs are currently used as additives for various applications such as high vacuum, extreme temperature and micro-electromechanical machines [40, 45]. Novel oil-miscible IL additive development can be a challenge; however, using IL as an emulsion may avoid solubility issues [46].

3.4 Types of tribological testing to evaluate the multifunctionality of lubricants

Tribology is the science of friction, wear, and lubrication between two surfaces that are in relative motion to each other. Proper lubricity in the contact region reduces friction and wear. Lubricants may react with contacting surfaces either physically or chemically, forming a protective surface layer called 'tribofilms' or 'tribolayers.' These layers reduce friction and wear between the abrasive surfaces. High lubricity means reducing mating part asperity contacts, resulting in decreased friction and energy loss [47]; however, high lubricity may not exhibit enhanced wear protection in all cases. Several researchers have conducted a range of tribological tests to evaluate a lubricant's friction and wear characteristics, some of which detail the critical multifunctionality of bio-based lubricants. A wide variety of experimental facilities were used in these tests; therefore, the most common types of test configurations used to evaluate the tribological performance of bio-based lubricants must be understood (figure 3.1). Figure 3.1(a) illustrates the four-ball tribotester, a versatile test method that conforms to ASTM D4172 [48]. The extreme pressure properties of bio-based lubricants can be evaluated using ASTM D2783 [49], where three of the four balls are fixed in a ball pot filled with bio-based lubricants. The fourth ball presses against the bottom three balls to conduct the test under specific

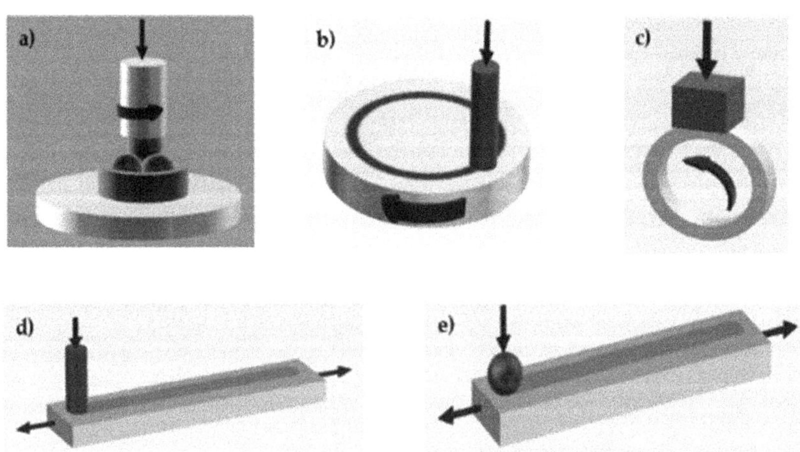

Figure 3.1. Widely used tribological test geometry configurations (a) four-ball, (b) pin-on-disc, (c) block-on-ring, (d) pin-on-flat, and (e) ball-on-flat tribometer. Reproduced from [61] CC BY 4.0.

conditions. The tribological performance of neem oil [50], rapeseed oil [51], jatropha oil [52], and pongamia oil [53] was investigated using this test configuration. Figure 3.1(b) demonstrates the popular pin-on-disc configuration, where a controlled load is applied through the pin on a disc that rotates at a specific angular velocity, according to the ASTM G99 standard. The tribological performance of several bio-based lubricants, such as rapeseed oil [54], polanga oil [55], jathropa oil [13], and *Phyllanthus emblica* seed oil [56] was evaluated with this apparatus. Figure 3.1(c) presents the block-on-ring type, where the block, or top part, presses against the rotating ring at controlled test conditions to simulate bearings or clutch facing cases, in accordance with ASTM G77. The multifunctionality of sunflower oil [57, 58] was tested using this configuration. The last two configurations, figures 3.1(d) and (e), are called pin-on-flat and ball-on-flat, respectively. The flat surface at the bottom reciprocates at a specific frequency in both cases. The top part is either a ball or a pin pressed against the flat under controlled pressure to evaluate the tribological behavior of the substance, according to ASTM G133. The tribological behaviors of methylesterified bio-oil [59], palm, and soybean oils [60] were tested using this configuration.

Some relatively uncommon experimental setups, along with the previous test configurations, were used to analyze the multifunctional behavior of bio-based lubricants and ILs. Tomasz Trzepiecinski (2020) studied the tribological performance of environmentally friendly edible vegetable oils in a forming operation using a strip drawing test [62] (figure 3.2(a)).

The test specimen was in the form of a sheet metal strip housed between some cylindrical counter surfaces (figure 3.2(b)). Talib and Rahim (2016) investigated the tribological performance of modified jatropha oils as a metalworking fluid [63]. An orthogonal cutting experimental setup was used to evaluate the performance of

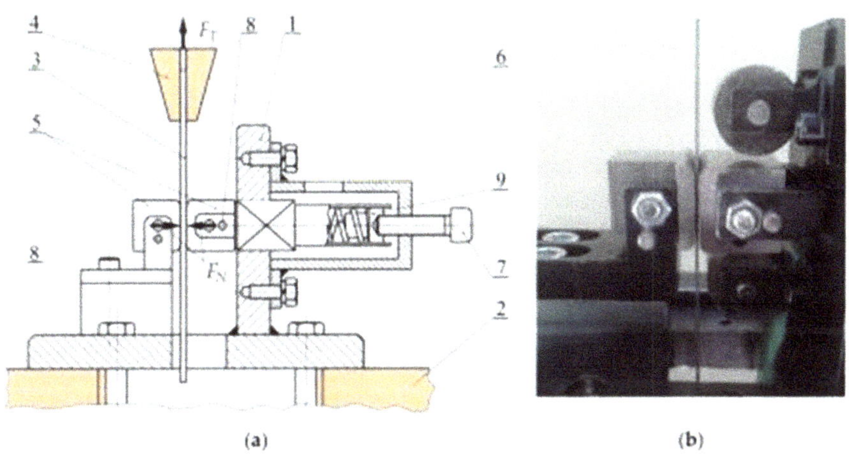

Figure 3.2. (a) Schematic of the experimental setup and (b) an image of the working zone: 1—frame, 2—bottom grip of the testing machine, 3—specimen, 4—the upper grip of the testing machine, 5—counter samples, 6—spring, 7—bolt, 8—fixing pins, 9—Teflon washer. Reproduced from [62] CC BY 4.0.

Figure 3.3. Orthogonal cutting experimental setup. Reprinted from [63], copyright (2016), with permission from Elsevier.

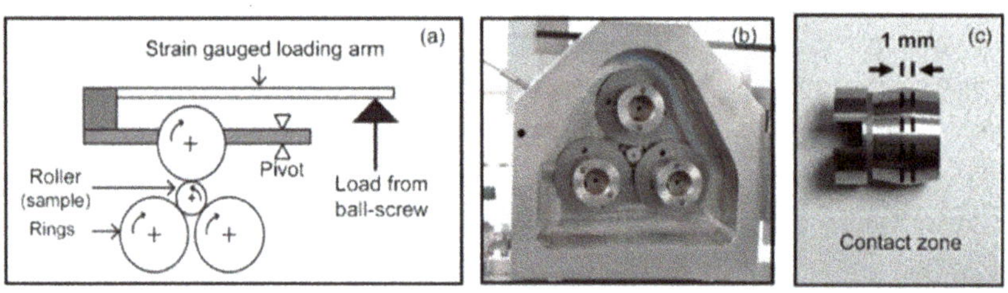

Figure 3.4. (a) Schematic diagram of three-disc on a roller-type rolling contact fatigue test rig (b) image of test chamber (c) image of 1 mm face-width chamfered test sample. Reprinted from [66], copyright (2018), with permission from Elsevier.

bio-based oils (figure 3.3). This apparatus used a thermal imaging camera, a dynamometer, and a nozzle that directed the oils into the machining zone.

ILs that can be used as oil-miscible additives for rear axle lubrication in automobiles were recently studied by Roy *et al* (2021) [64, 65]. Three discs on roller-type rolling contact fatigue test rigs were used to simulate the planetary gear train observed in rear axles (figure 3.3(a)). IL additized mineral oils were tested with a dip lubrication system to evaluate their performance against that of commercial rear axle oils. The test rig sample chamber and test sample (roller) containing a 1 mm wide contact region are illustrated in figures 3.4(b) and (c), respectively.

3.5 Tribological behavior: friction behavior, wear behavior

Many peer-reviewed journal articles have reported on the tribological characteristics of natural vegetable oils, chemically modified vegetable oils, and nanoparticle additized or IL additized oils. A set of key findings on each of these domains is discussed in sections 3.5.1–3.5.3.

3.5.1 Tribological characteristics of pure vegetable oils

Researchers have investigated the physiochemical characteristics and tribological behaviors of a wide variety of plant-based lubricants, such as soybean, rapeseed, jatropha, canola, palm, and sunflower oils. Soybeans are the most common base stock in the USA since the resource is abundant. Rapeseed is the most widely studied bio-based lubricant in Europe since it is similarly abundant. Soybean farming is advantageous because humans can consume the product; however, rapeseed is edible and can produce more oil per unit area than many other oil sources.

Kraipat Cheenkachorn (2013) studied various parameters affecting the friction and wear resistance of several soybean oils: conventional soybean oil (NSB), high-oleic soybean oil (HSB), and epoxidized soybean oil (ESB) [67]. Table 3.3 lists the fatty acid distribution by weight percentage.

Epoxidized soybean oil is the conventional soybean oil, where the double bonds were epoxidized to form epoxide rings. The commercially available anti-wear additive (AW) zinc dithiophosphate (ZDDP) was added at 1% to enhance its tribological behavior. Tribological tests were performed using a 25 lb load and AISI 52100 balls at three speeds (600, 900, and 1200 rpm) and three different operating temperatures (25 °C, 50 °C, and 75 °C). Figures 3.5(a)–(c) display the friction coefficients for three oils and ZDDP mixtures at velocities of 600, 900, and 1200 RPMs, respectively. Friction data from these three oil types did not indicate any clear trend. The viscosity of the ESB oil type is higher than the NSB and HSB types at room temperature, but the viscosity of ESB decreases at increased temperatures. ESB without ZDDP had the highest friction coefficient at 25 °C.

Increasing the speed resulted in an increased sliding distance, increasing the wear scar diameter. ESB had a higher viscosity than the other two lubricants at the same temperature. The continuous generation of a high-viscosity layer and the potential development of hydrodynamic effect occurred at low speeds. The adsorption of the epoxide ring in the alkyl chain was strong enough for surface protection at 25 °C and 50 °C without the addition of ZDDP; however, the wear protection was not significant due to the removal of oil film at temperatures above 50 °C. Electron micrographs or chemical analyses were not presented.

Habibullah *et al* (2014) studied the effect of additizing 1–5 vol%, Jatropha oil to SAE 40 and determined its tribological behavior on steel [52]. The authors observed that the addition of 5 vol% Jatropha oil resulted in the best anti-wear properties

Table 3.3. Fatty acid distribution in experimented soybean oils.

	Palmitic acid	Stearic acid	Oleic acid	Linoleic acid	Linolenic acid
NSB	10.5	4.1	23.4	52.6	7.2
HSB	7	4	83	3	2

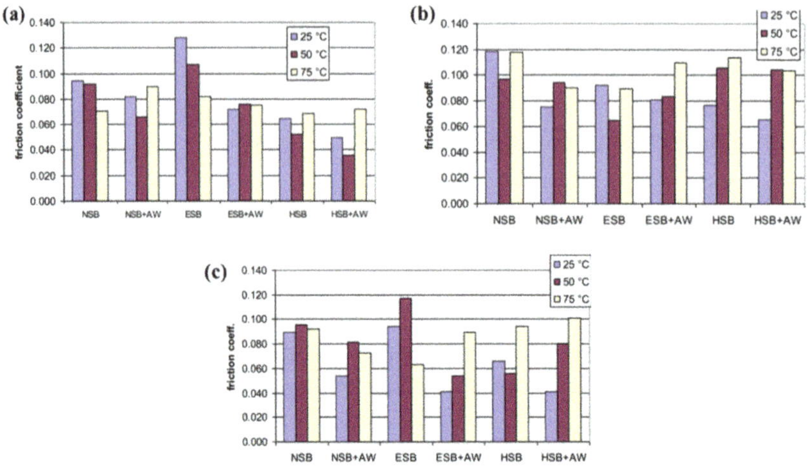

Figure 3.5. Friction coefficients for different soybean oil-based lubricants at (a) 600, (b) 900, and (c) 1200 rpm. Reprinted from [67], copyright (2013), with permission from Elsevier.

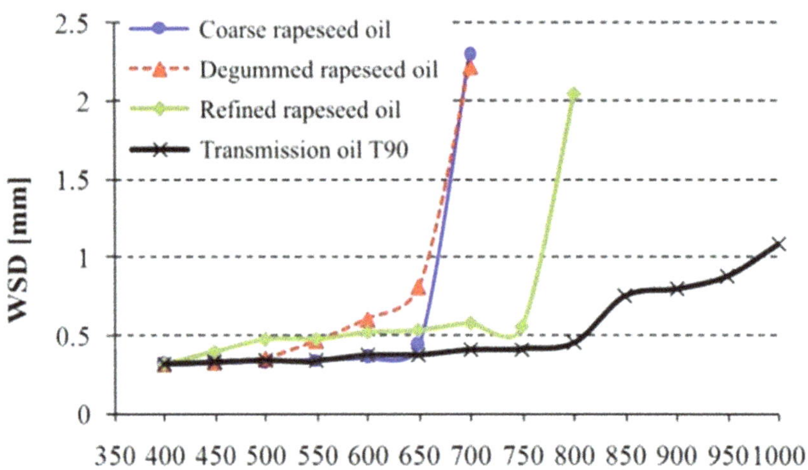

Figure 3.6. Evolution of wear scar diameter (WSD) with load for different vegetable oil lubricated conditions. Reproduced from [68], copyright IOP Publishing Ltd. All rights reserved.

amongst the blend versions due to a higher kinematic viscosity value. Various versions of rapeseed oils were tested using a four-ball tester then compared to transmission oil T90 [68], similar to the non-additized soybean oil tests. The critical challenges observed for rapeseed oils are a lower load limit to seizure and a steeper gradient to seizure limit (figure 3.6).

Vegetable oils are sensitive to oxidation at higher temperatures, which is a significant disadvantage. Mannekote and Kailas (2012) studied five different bio-based oils: groundnut, palm, rice bran, soybean, and sesame [24]. These oils were stored at elevated temperatures in the dark to avoid photo-oxidation. Accelerated

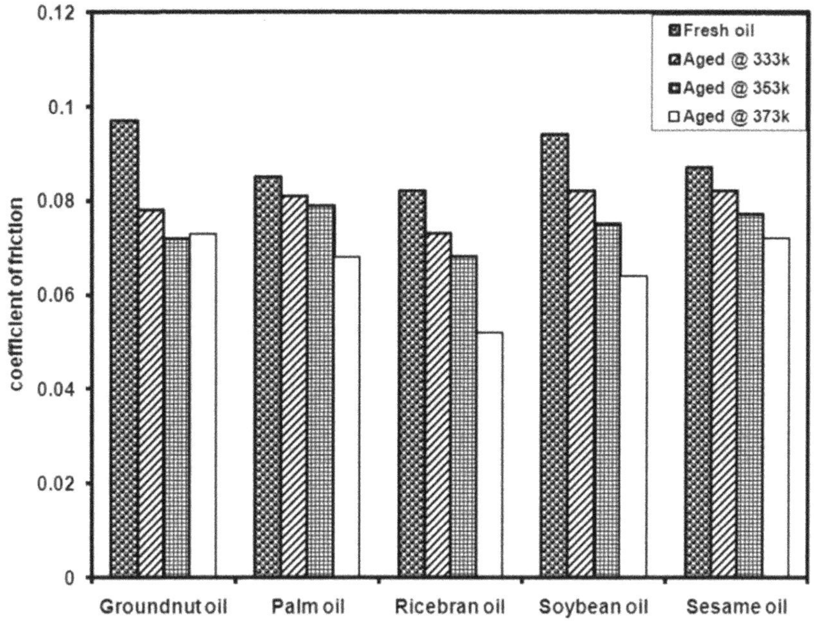

Figure 3.7. Variation of friction coefficient with different vegetable oil aging temperatures. Reprinted from [24], copyright (2012), with permission from Elsevier.

aging and the effect on tribological behavior were investigated using a four-ball tester. Figure 3.7 illustrates the frictional characteristics of these aged oils at different temperatures compared to fresh oil. The reduced friction of aged and higher temperature oils is attributed to the formation of soft iron soap with a low shear resistance.

3.5.2 Tribological characteristics of additized vegetable oils

Researchers have studied a range of different additives to mitigate the challenges of neat vegetable oils. These additives have specific properties, such as dispersant, detergent, anti-oxidation, anti-corrosion, anti-foam, viscosity modification, anti-wear, and pour point depression; however, some details must be considered when additizing vegetable oils.

- More than one type of additive package needs to be used, and synergistic or antagonistic behaviors may be observed on their tribological characteristics.
- Additives should be ashless and non-toxic [69]. The ashless metals Ca, Na, K, and Mg, and Ca-based soaps may be used, which are less toxic than Li-based soaps.
- The agglomeration and aggregation of nanoparticles in the vegetable oil solvents.
- The physical and chemical stability of the oils with regard to application temperature and time.

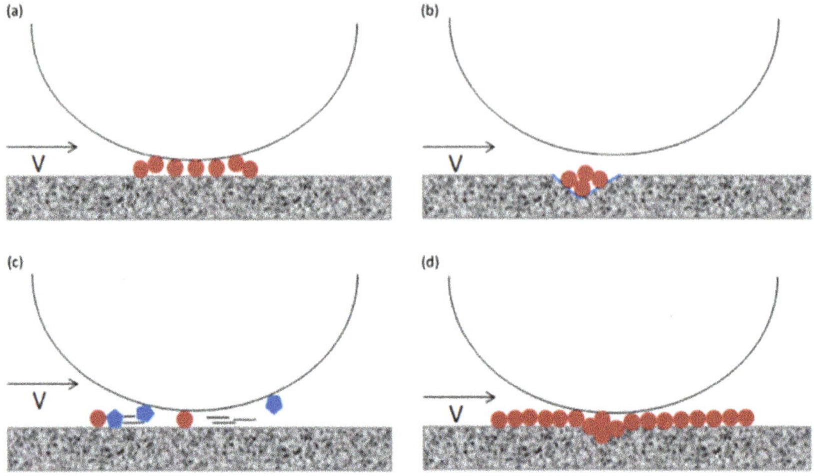

Figure 3.8. Various lubricating mechanisms of water-based lubricants with TiO_2 nanoparticles as an additive (a) rolling effect, (b) mending effect, (c) polishing effect, and (d) protective film effect. Reprinted from [70], copyright (2017), with permission from Elsevier.

Several hypotheses have been presented on the lubrication mechanisms that different nanoparticles offer when applied to contact regions. Wu *et al* (2016) discussed the four fundamental mechanisms for TiO_2 nanoparticles in water-based lubricants: rolling, mending, polishing, and protective film formation (figure 3.8) [70]. Other researchers have studied these four mechanisms while analyzing the applicability of nanoparticles in the tribological contact zones. Contact pairs are separated slightly due to the rolling effect [71–75] and protective tribofilm growth [76–79], which reduces the friction coefficient of the tribopair. The other two mechanisms, mending [80] and polishing [71, 81], enhance the surface quality and reduce surface damage during frictional contacts.

One key challenge in using nanoparticles is determining if they react with the solid surfaces or the solvent oil properly through physical or chemical reactions. Researchers have used broadly seven types of nanoparticle-based friction and wear modifiers:

- Carbon material based NPs: diamond powders [82, 83], graphite [84], fullerene [85], [86], carbon black [87], graphene oxide [88], carbon nanotubes [89, 90].
- Metals: Cu [26, 91], Ag [92], Co [93], [94].
- Metal oxides: CuO [75], [95–97], TiO_2 [82, 98, 99], FeO [100], ZrO_2 [101], SiO_2 [95], [102].
- Solid lubricants: WS_2 [103], hBN [104], MoS_2 [105].
- Metallic salts: $CaCO_3$ [106], Zr phosphate [107], hydrosilicate [108].
- Rare earth elements: La borate [109], $La(OH)_3$ [110].
- Organic: tri (hydroxymethyl) propane ester [111], PTFE [112].

Table 3.4 lists some representative friction and wear modifiers used to enhance tribological behavior in different contact conditions.

Table 3.4. Use of nanoparticle-based additives to enhance tribological performance.

Additive type	Base lubricant	Testing method	Key findings	References
CuO and ZnO	Soybean and sunflower via epoxidation reaction	High-frequency reciprocating rig (HFRR)	The addition of NPs did not enhance tribological behavior. This lack of enhancement may be due to the high adsorption ability of bio-lubricant on a metal surface and the NPs resultant third body wear.	[100]
CuO and Al_2O_3	PAO8 and GL4 (SAE 75W-85)	Four-ball tribotester	Increased protection against wear was observed by CuO NPs. Wear protection was higher for PAO base oil compared to GL4. Al_2O_3 NPs acted as abrasives and did not result in any surface protection.	[97]
Cu	SF15W/40	Ball-on-disc	Oil-soluble Cu NPs can show long-term stability in SF15W/40 oil. 0.8 mass% NP concentration showed the best tribological behavior.	[91]
Cu and serpentine mixture	Commercial diesel engine oil (CD 15W/40)	Ball-on-disc	The most effective synergistic effect on tribological behavior was observed with a 7.5 wt% Cu NP composition.	[113]
ZnO	Commercial base oil (specific type was not mentioned)	Four-ball tester	ZnO NPs reduced wear with the formation of a tribofilm on the contact surface. The wear scar diameter was lowest with 3 wt% ZnO concentration.	[114]
CuO and WC	PAO	Three discs on roller	CuO performed slightly better than PAO to control micro pitting and wear, but the WC NPs based lubricant enhanced the rolling contact fatigue life of the samples significantly.	[75]
$ZnAl_2O_4$	Pure lubricating oil (specific type was not mentioned)	Four-ball tester and thrust ring tester	$ZnAl_2O_4$ NPs with 0.1 wt% concentration exhibited better friction and wear response compared to Al_2O_3 and ZnO NPs.	[115]
Ag	Commercial multiply alkylated cyclopentane (MAC)	Oscillating ball-on- disc test rig	A 2 wt% addition of Ag NPs improved wear performance and load-carrying capacity significantly. Minimal benefit on friction reduction was observed.	[92]

Nano graphene	Soybean oil	Four-ball tester	The addition of graphene did not improve the coefficient of friction. No significant benefit in wear resistance was observed due to increasing the graphene percentage from 0.25 to 1 wt%.	[116]
TiO_2	Water-based lubricants	Ball-on-disc	The addition of TiO_2 reduced the friction coefficient and ball wear by 49.5% and 97.8%, respectively	[70]
Nano PTFE	150 N API Group II base oil	Four-ball tribotester	3 wt% PTFE presented the best tribological performance due to tribofilm formation. Extreme pressure property could be enhanced by 454%, making nano PTFE a promising anti-wear and anti-friction compound.	[112]
Graphene oxide (GO)	Rapeseed oil	Four-ball tribotester	Two versions of graphene oxide-based additives (GO-D and GO-T) presented enhanced tribological behavior compared to a GO. 0.2 wt% of GO-D, resulting in a 40.1% reduction in wear scar diameter.	[88]

Next-generation researchers that study bio-based lubricants should consider that the addition of nanoparticles cannot always provide enhanced friction and wear. Trajano *et al* (2014) [100] investigated the tribological effects of metal oxide nanoparticle additions (ZnO and CuO) to soybean and sunflower oils under boundary lubrication. A high-frequency ball-on-disc type reciprocating rig was used to test a formulation of lubricant and 0.5% nanoparticles by weight based on previous literature [117, 118]. The soybean and sunflower oils were both epoxidized before the addition of NPs, and the tribological behavior of the NP additized oils were compared to epoxidized base oils. The epoxidized pure vegetable oils had a lower friction coefficient than that of nanoparticle added epoxidized soybean and sunflower oils. The epoxidized vegetable oils exhibited enhanced lubricity due to their remarkable viscosity index over a wide temperature range, and improved film formation at the contact region. The notable bio-lubricant adsorptibility into the metal surface hindered nanoparticle deposition resulting in the nanoparticles acting as third body abrasive particles instead of lubricant.

Cristea *et al* (2018) added carbon-based material nano graphene to soybean oil then analyzed its tribological behavior [116]. Three different concentrations

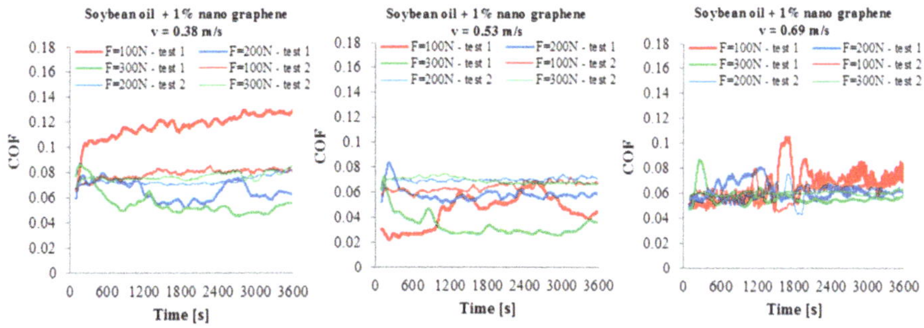

Figure 3.9. The evolution of the coefficient of friction (COF) with time as a function of load and speed with soybean oil and 1 wt% nano graphene. Reproduced from [116], copyright IOP Publishing Ltd. All rights reserved.

(0.25 wt%, 0.5 wt%, and 1 wt%) were mixed with soybean oil then tested with three different loads (100 N, 200 N, and 300 N) and three different sliding velocities (0.38 m s^{-1}, 0.53 m s^{-1}, and 0.69 m s^{-1}) using a four-ball tester. Two levels of friction coefficients were observed at a sliding velocity of 0.38 m s^{-1} (figure 3.9). The authors hypothesized that EHD lubrication resulted in a lower friction coefficient, below 0.1, and a mixed or boundary lubrication resulted in a higher friction coefficient, above 0.1; however, the addition of graphene did not significantly improve the reduction in the friction coefficient for all three sliding conditions.

Zhang *et al* (2018) synthesized two types of lubricant additives derived from graphene oxide (GO) with 1-dodecanethiol (GO-D) and tert-dodecyl mercaptan (GO-T) reactions [88]. These additives were later added to rapeseed oil, and the tribological behavior was investigated using a four-ball tribotester. A 0.2 wt% addition of GO-derived additives presented the lowest friction coefficient for a wide range of loading conditions (98 N–392 N). The GO-derived additive mixed formulations exhibited a lower friction coefficient than standalone GO in all cases (figure 3.10). Similar observations were made for wear resistance.

The tribological behavior of 0.2 wt% additized GO, GO-D, and GO-T was tested and compared to non-additized rapeseed oil. Figures 3.11(a) and (b) illustrate the friction and wear behavior of those four lubricants as a function of varied loads between 100 N and 400 N. The friction coefficient and wear scar diameters were comparable for both 0.2 wt% GO and pure rapeseed oil. The addition of GO-D and GO-T reduced the friction coefficient and wear scar diameter. The tribological behavior enhancement increased with an increased contact load. GO-D and GO-T exhibited remarkable anti-wear and anti-friction properties due to the layered structure of these additives and a stable dispersion in rapeseed oil.

3.5.3 Tribological characteristics of bio-based ionic liquids

3.5.3.1 Ionic liquids—their benefits and challenges in application

ILs are room-temperature molten salts consisting of cations and anions. ILs possess unique physiochemical characteristics that indicate their remarkable suitability as lubricating candidates, such as low volatility, melting point, inflammability, and high

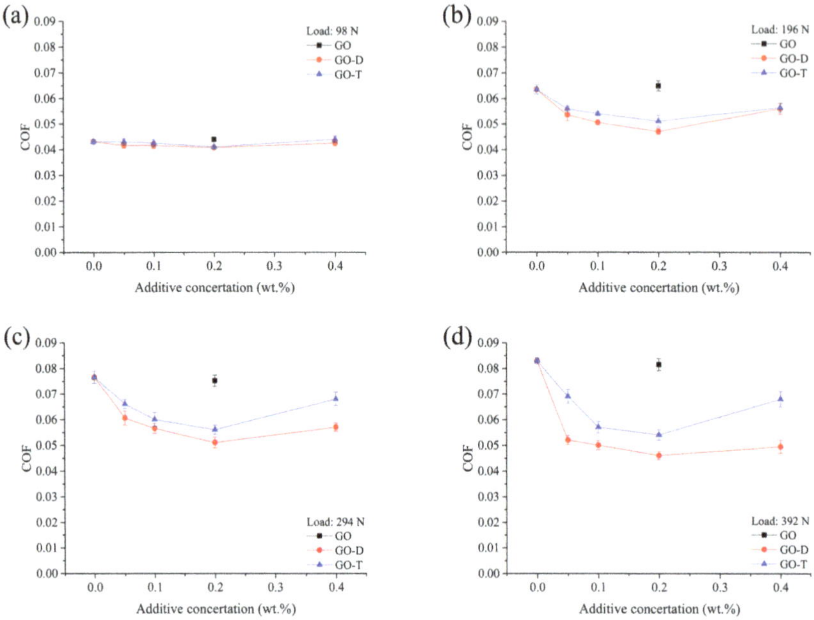

Figure 3.10. COF with additive concentration as a function of load and lubricant types: GO, GO-D, and GO-T. Reprinted from [88], copyright (2018), with permission from Elsevier.

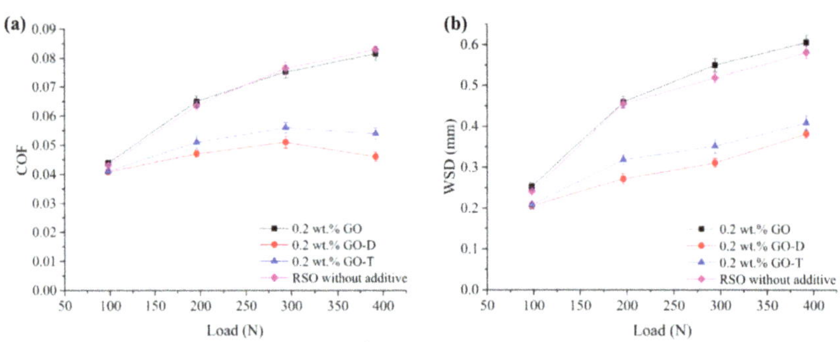

Figure 3.11. (a) COF and (b) wear scar diameter (WSD) with load for 0.2% GO, GO-D, GO-T, and rapeseed oil without additive. Reprinted from [88], copyright (2018), with permission from Elsevier.

thermal stability [119]. ILs share several properties with traditional bio-lubricants, such as polar molecules, derivation from bio-based stock, lamellar structure, and remarkable lubricity [119–122]. Oils containing polar groups, such as esters and carboxylic acids, can efficiently adsorb and react with metal surfaces [123], which aids in the formation of a stable tribofilm in boundary lubrication and results in reduced friction, vibration, and wear. The properties of tribofilms may vary depending on the cation or anion types and how they react with the solid surface. Metal surface corrosion, poor miscibility or stability in the base oil, low hydrolytic stability, and high production cost may be observed when ILs are introduced [119, 124].

Tribological properties from a range of ILs have been tested on different metal surfaces by several researchers. A large number of studies have been conducted on steel or ferrous alloys since most machinery is built using these alloys [64, 65, 125–132]. IL additized lubricants exhibit enhanced tribological behavior on Al alloys [133–137], Cu alloys [37, 125, 138–140], Ni-based alloys [140–142], and polymers [143–146]; however, not all ILs are environmentally benign. Tetrafluoroborate (BF_4), hexafluorophosphate (PF_6) or bis(trifluoromethylsulfonyl)imide (TFSI) anions can produce fluorine-based compounds via hydrolysis. This hydrolysis results in a corrosive environment that is toxic and must be carefully monitored [147–150]. Similar issues were observed in the case of sulfur and phosphorus-containing ILs [151]. Jimenez *et al* (2006) demonstrated that imidazolium ILs containing BF_4 and PF_6 anions result in tribochemical reactions at aluminum–steel contacts [152]. Freire *et al* (2010) [149] analyzed that such reactions can occur under certain pH levels and temperatures. A significant effort was led to develop stable fluorine-containing anions, such as FAP and NTf_2, to avoid these corrosion issues [153–158].

3.5.3.2 Environmentally friendly ILs and their tribological behavior

Some research is currently being performed to develop environmentally friendly ILs without halogens and metals [32, 159–161]. A new set of ILs are being developed with reduced toxicity and enhanced biodegradability [13, 46, 162, 163]. The first fatty acid anion-based ILs were reported in 2013 [164], and many studies have been conducted to analyze their tribological behaviors [165–174]. Gusain and Khatri (2016) reported a 20%–50% reduction in the COF due to the addition of fatty acid-based ILs compared to a polyol ester base oil [167]. Faes *et al* (2021) developed three new fatty acid anion-based environment friendly ILs (FAILs) that were synthesized from natural sources [166]: methyltrioctylammonium hexanoate ($[N_{8881}][_{C6:0}]$), methyltrioctylammonium octadecanoate ($[N_{8881}][C_{18}:0]$), and methyltrioctylammonium octadec-9-enoate ($[N_{8881}][C18:1]$). The tribological behavior of these three ILs was compared to five different tribopairs using a ball-on-disc type tribometer. Figure 3.12 displays the frictional behavior in five different tribopairs with the three IL lubricated conditions. The frictional behavior of these three ILs was similar, with the lowest COF for aluminum and the highest COF for WC materials, which can be correlated to the change in contact pressure due to the change in the contacting material's elastic modulus.

Table 3.5 lists the wear volume ($\times 10^6$ μm^3) numbers for different tribological tests using the three ILs. The lowest wear was observed for WC samples due to their increased hardness. Increased hardness was also cited for the higher wear volumes of bronze and aluminum.

Reeves *et al* (2017) synthesized anions from artificial sweeteners such as saccharinate, and food additives such as salicylate and benzoate to develop environmentally friendly ILs [175]. Cations, such as trihexyltetradecylphosphonium salts (P_{66614}), are compatible with the previously mentioned anions and exhibit antimicrobial properties [175]. The room temperature moiety structure of the experimented ILs is illustrated in figure 3.13.

Pin-on-disc tribological tests were performed using three bio-based environmentally friendly IL additized lubricants, salicylate, saccharinate, and benzoate,

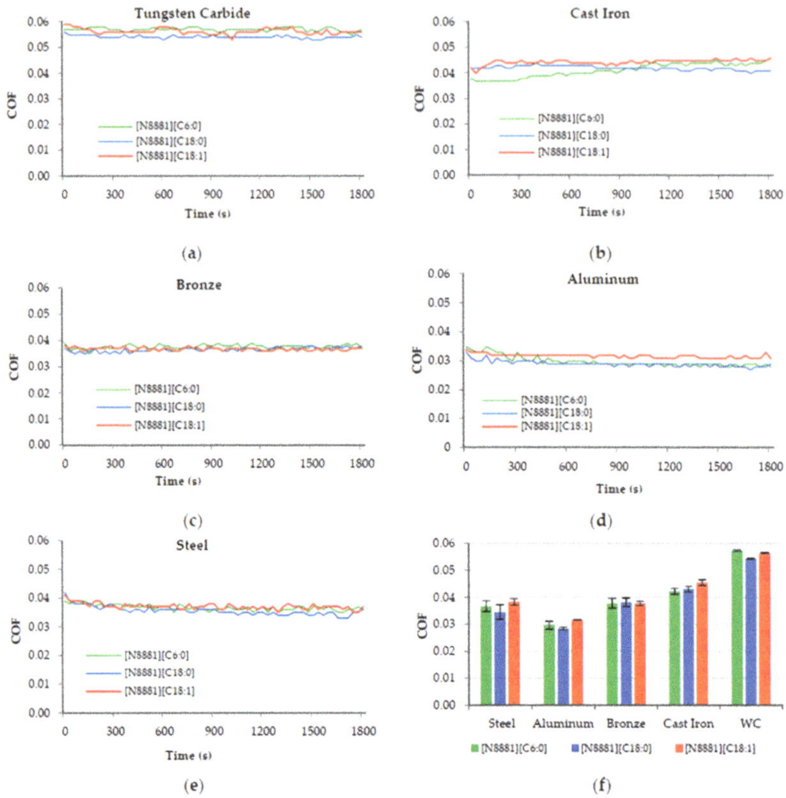

Figure 3.12. COF with time on (a) tungsten carbide, (b) cast iron, (c) bronze, (d) aluminum, (e) steel, and (f) average COF on all surfaces. Reprinted from [166] CC BY 4.0.

Table 3.5. Wear volume (× 10^6 μm^3) on different surface types lubricated using three experiments ILs. Reprinted from [166] CC BY 4.0.

FAIL	$[N_{8,8,8,1}][C_{6:0}]$		$[N_{8,8,8,1}][C_{18:0}]$		$[N_{8,8,8,1}][C_{18:1}]$	
Surface	Average value	Standard deviation	Average value	Standard deviation	Average value	Standard deviation
WC	0.275 168	0.009 613	0.078 149	0.009 957	0.214 120	0.010 297
Cast Iron	9.205 816	0.018 750	8.198 145	1.045 135	7.871 593	1.150 991
Bronze	8.957 983	0.774 445	9.277 544	0.967 261	7 536 845	0.834 904
Aluminum	8.235 090	0.557 433	6.949 583	0.002 316	7.763 727	0.037 912
Steel	6.472 298	0.146 255	6.558 970	0.075 222	7.036 670	0.086 427

and a less environmentally friendly IL, 1-decyl-3-methylimidazolium bis trifluoromethyl sulfonylimide ($C_{10}mimTf_2N$). The tribological behavior of these ILs were compared to petroleum-based and natural vegetable oils. Figure 3.14 displays the friction coefficient numbers for Al-steel contacts under a constant load of 10 N and a

	Name and structure	
Cations	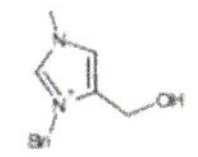 1,3-dialkylimidazolium	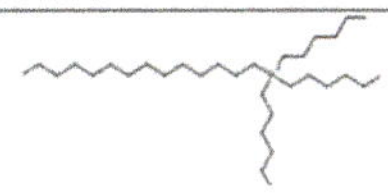 trihexyltetradecylphosphonium ($P_{666,14}$)
Anions	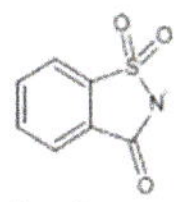 Saccharinate	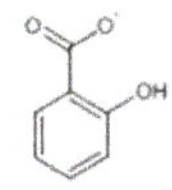 Salicylate

Figure 3.13. Moiety structure of room temperature ILs. Reprinted by permission from *Journal of Bio- and Tribo-Corrosion*: Springer Nature, copyright (2017) [175].

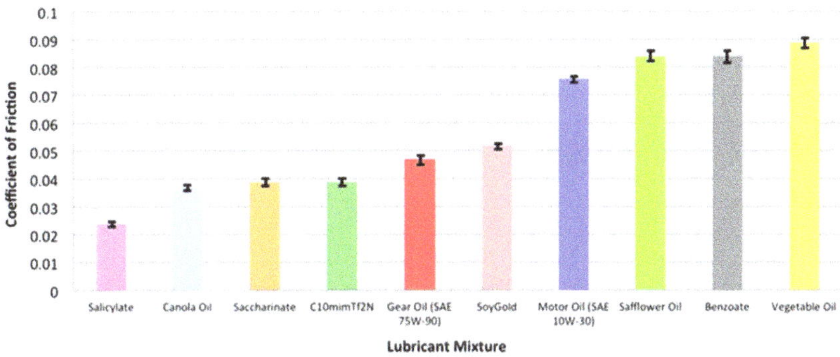

Figure 3.14. Variations of the COF under different lubricated conditions. Reprinted by permission from *Journal of Bio- and Tribo-Corrosion*: Springer Nature, copyright (2017) [175].

linear velocity of 36 mm s^{-1}. The formulated ILs exhibited a lower friction coefficient than many of the petroleum-based and natural oils. Higher absorptivity due to unique dipolar structures was attributed to this enhanced IL frictional behavior.

ILs exhibited increased wear resistance compared to petroleum-based and natural oils (figure 3.15). Salicylate and saccharinate functional groups, along with the alkyl chain length of IL lubricants, played a critical role in enhancing the anti-wear properties of the lubricants. The formation of an anion-based monolayer film onto a positively charged metal surface increased the surface protection in boundary lubrication.

Jiang *et al* (2018) developed Cholin (Ch) cation-based ILs and investigated their tribological properties [161]. Choline is a water-soluble nutrient recognized as a B-complex vitamin. Choline-based ILs are known for their low toxicity and reasonable biodegradability [161]. Amino acids (AAs) are popular, cost-effective biomaterials that are readily available in nature. [Ch][AA] based ILs are devoid of halogen, sulfur, and phosphorus, making them environmentally friendly, biodegradable, sustainable, and non-toxic [10, 176]. Jiang *et al* (2017) synthesized five [Ch][AA] based ILs with an identical cation (Cholin) but different anions [161]. The effect of alkyl chain length and anion symmetry on tribological behavior was investigated using oscillating reciprocating friction and a wear test rig. The five ILs formulated for this study were choline glycine ([Ch][Gly]), choline alanine ([Ch][Ala]), choline isoleucine ([Ch][Ile]), choline phenylalanine ([Ch][Phe]), and choline serine ([Ch][Ser]). Figure 3.16 illustrates the frictional behavior of these lubricants at room temperature. [Ch][Gly] IL exhibited the lowest friction coefficient, which was 20% lower than the traditional IL [C_6mim][NTf_2]. Increasing the ILs anion size increased the viscosity; therefore, most of the formulated [Ch][AA] based ILs presented a reduction in friction compared to traditional IL-based lubricant without adding a friction modifier. The wear resistance of the [Ch][Gly] lubricated condition was lower than [C_6mim][NTf_2]. Low viscosity might be attributed to this inadequate wear resistance.

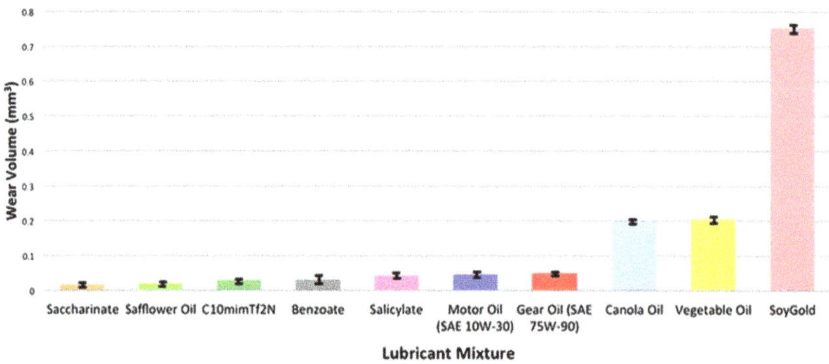

Figure 3.15. Variation of wear volume under different lubricated conditions. Reprinted by permission from *Journal of Bio- and Tribo-Corrosion*: Springer Nature, copyright (2017) [175].

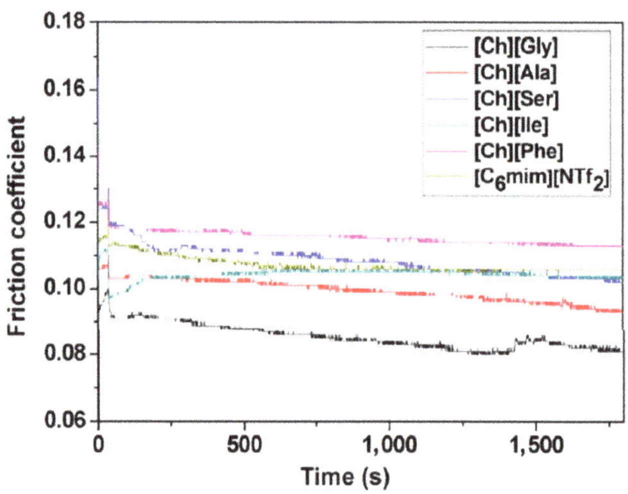

Figure 3.16. COF with time lubricated with [Ch][AA] ILs and [C_6mim][NTf_2]. Reprinted by permission from *Friction*: Springer Nature [161], copyright (2017).

3.5.3.3 Ionic liquids and their application in rolling-sliding contacts

Many of the tribological studies promoted the beneficial effect of IL additization in the context of sliding contacts. A set of studies have recently been conducted that investigate the impact of ILs as an additive to enhance rear axle lubrication, which can boost fuel economy in passenger vehicles [64, 65, 177]. The key ILs used in these studies were [P_{8888}] [DEHP], [N_{888}H] [DEHP], [P_{66614}] [$C_{17}H_{35}COO$], and [P_{66614}] [BTMPP]. Three discs on roller-type rolling contact fatigue tests in low and high sliding conditions were performed to simulate rear axle lubrication. These ILs contained P; however, the experimental results were promising. Stump *et al* (2019) presented the unique functionality of ILs to mitigate rolling contact fatigue (RCF) failures in gears and bearings [177]. These ILs were mixed with VHVI8 base oil and compared to the RCF behavior of conventional rear axle oil (75W-90). Figure 3.17 depicts the vibration signals (figure 3.17(a)), friction (figure 3.17(b)), and wear volume (figure 3.17(c)) results captured during the RCF tests using [P_{8888}] [DEHP] IL (IL1) lubricated condition and conventional oil (75W-90). The viscosity of the IL additized oil was half of the conventional lubricant; however, it outperformed the conventional lubricant in all three domains: vibration, friction, and wear. Scanning tunneling electron microscope (STEM) imaging on focused ion beam (FIB) cross-sectioned samples from sample surfaces confirmed the formation of a tribolayer that protected the surface from RCF failure modes such as crack formation and micro pitting.

Roy *et al* (2021) developed fully formulated oils, including different additive packages along with phosphonium phosphate ([P_{8888}] [DEHP]) IL [64]. A range of IL additized formulations were prepared since antagonistic behavior between different additive packages or between an additive package and IL can be experienced. Detailed experimental results of three fully formulated oils were presented and compared to conventional fully formulated oil (SAE 75W-90). The fully formulated oils had an approximate viscosity that was half of conventional 75W-90; however, the friction, wear, and vibration signals for a couple of IL additized oils (FF V1 and FFV2) were comparable. Figure 3.18 illustrates that the surface protection offered by IL additized fully formulated oils was significantly better than that of 75W-90. The surface of the 75W-90 lubricated roller surface was significantly cracked and micropitted in both low and high slide to roll ratio (SRR) conditions, unlike that of the fully formulated oils. The spacing between the cracks was increased at high sliding due to a change in the rotational velocity of the sample.

Figure 3.19 presents the elemental mapping and STEM images of FIB lifted work surface cross-sections from the sample lubricated with IL additized fully formulated oil. The tribofilm thickness was 20–50 nm in the IL lubricated condition. The tribolayer thickness for conventional 75W-90 oil was 35–75 nm thick, confirming that the tribofilm thickness is not the most critical parameter needed to predict surface protection. Hub dynamometer tests showed 3.4% increased power output and 3.3% increased torque using IL containing test oil as compared to those of conventional gear oil.

Roy *et al* (2021) presented a unique IL additized lubricant behavior in the context of rear axle lubrication [65]. The authors investigated the surface protection

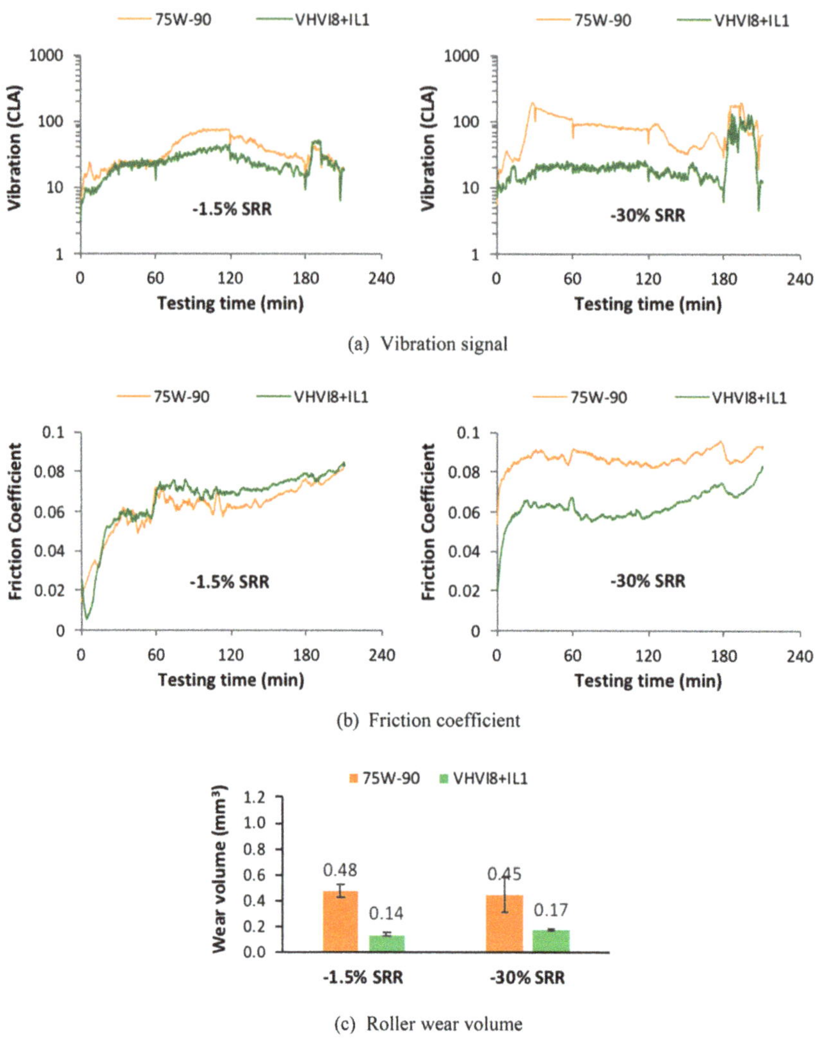

Figure 3.17. Comparison of (a) vibration signals, (b) COF, and (c) wear volume in low and high sliding conditions lubricated with conventional gear oil 75W-90 and IL additized lubricants. Reprinted with permission from [177], copyright 2019 American Chemical Society.

behavior of phosphonium phosphate ([P_{66614}] [BTMPP]) IL compared to a base oil (VHVI8) that did not contain IL. Both lubricants were tested using low and high sliding conditions. Figure 3.20 indicates that in both sliding conditions, increased wear resistance and surface protection were exhibited in IL lubricated cases. The surface protection remained significantly prevalent for both carburized AISI 8620 samples and through hardened AISI 52100 steel samples.

Interestingly, in spite of presenting increased surface protection, the vibrations for the IL additized cases were significantly higher than neat base oil in low sliding

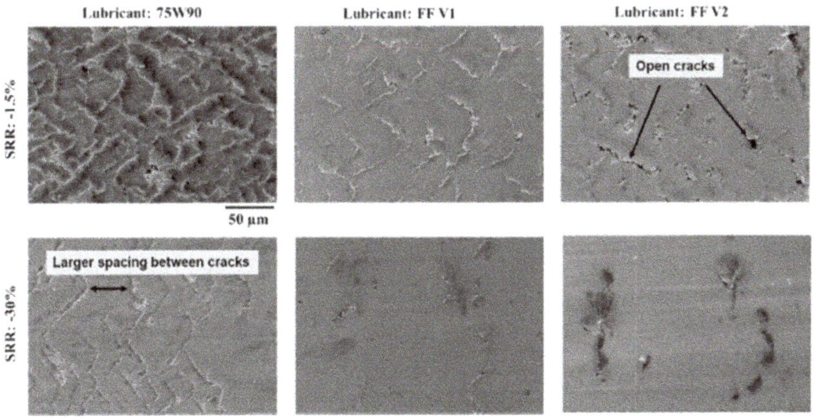

Figure 3.18. Comparison of sample surface topographies lubricated using conventional gear oil SAE 75W-90 and IL additized fully formulated oils. Reprinted from [64], copyright (2020), with permission from Elsevier.

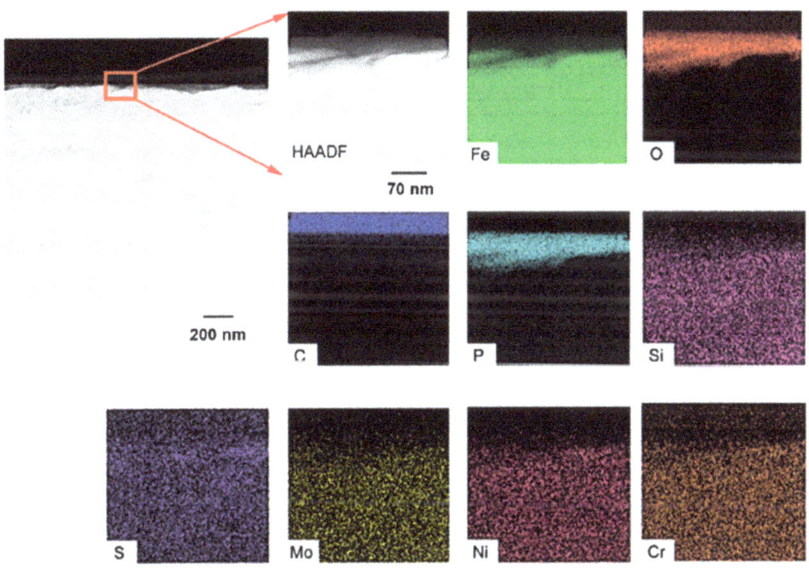

Figure 3.19. Cross-sectional STEM-EDS based tribofilm characterization on the sample lubricated using fully formulated IL additized lubricant under low sliding condition. Reprinted from [64], copyright (2020), with permission from Elsevier.

conditions (1.5% SRR) (figure 3.21). However, this behavior was not observed in high sliding conditions. Two sample materials (AISI 8620 steel and AISI 52100 steel) were studied to investigate whether the vibration phenomena are lubricant driven or material driven. A similar vibration response confirmed that the increased vibration in low SRR was caused by IL in the lubricant.

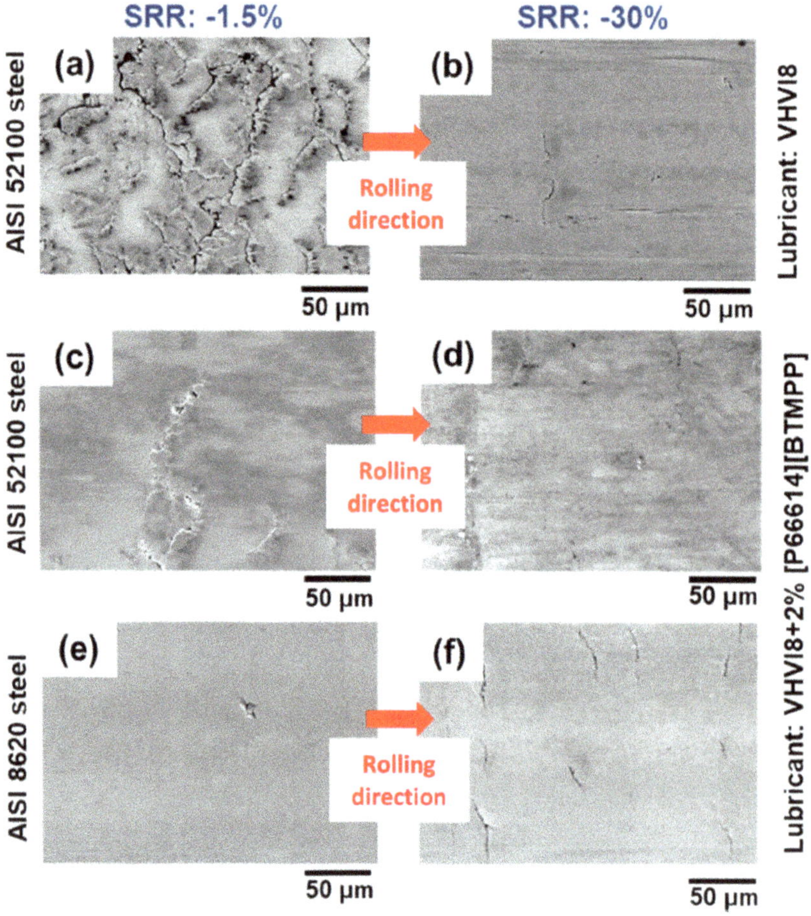

Figure 3.20. Scanning electron micrograph of worn rollers lubricated using VHVI8 base oil and VHVI8+IL under low (a, c, e) and high (b, d, f) sliding conditions. Reprinted from [65], copyright (2021), with permission from Elsevier.

Further tests were completed by intermittently stopping and restarting the analysis, which revealed the formation of several high-low regions and their evolution as the number of cycles was increased. These formations were caused by the chemical reaction between the IL and the sample surface. These tests also revealed an abnormal increase in vibration characteristics under the lubricated condition. Figures 3.22(a)–(f) illustrate the evolution of the test sample surfaces during the RCF experiments due to chemical reactions and mechanical fatigue. Figures 3.22(g) and (h) display a representative optical micrograph of the sample surface and its corresponding height map, respectively. These images demonstrate the formation of strip-like bands in the longitudinal direction, or along the rolling direction of the samples. The maximum depth of the height zones was 1 µm.

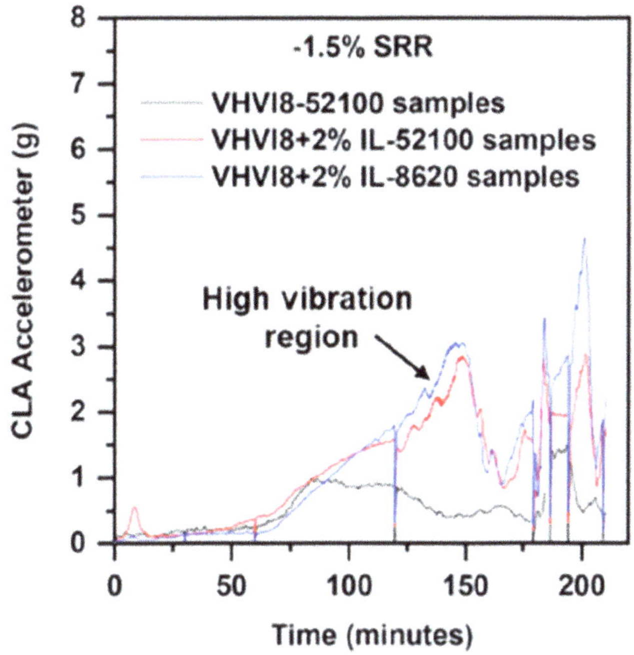

Figure 3.21. Vibration signals of RCF tests in low sliding lubricated using only base oil and base oil + IL lubricants. Reprinted from [65], copyright (2021), with permission from Elsevier.

3.6 Summary and future research perspectives

A detailed outlook of bio-based multifunctional lubricants and their tribological behavior are discussed in this chapter. The discussion has been divided into three categories, focusing on the characteristics of vegetable oil-based lubricants, nanoparticle additized lubricants, and IL additized lubricants. Increasing demands for energy will result in a scarcity of available lubricants if bio-based lubricant mass production and implementation are not thoroughly considered. An increased interest in ongoing research has been spurred by discussions of significant environmental changes and the depletion of fossil fuels. The renewability and biodegradability of existing and new bio-based vegetable oils must be investigated before they can be appreciably substituted for petroleum-based lubricants. The fatty acids found in vegetable oils are crucial components that determine the flashpoint, oxidative stability, pour point, and viscosity of bio-based lubricants. Applications for bio-based lubricants have wide varieties including engine oils, due to their low volatile organic compound emission, metalworking fluids, due to their substantial emulsifiability and other industrial sectors. Future research in this area should focus on exploring proper catalysts and cheaper feedstocks through the genetic modification of bio-based vegetable oils. Systematic research on lubricant preparation from bio-based feedstock, blending, and production methods should be encouraged. Research encompassing the performance of bio-based oils on real-world machinery, as

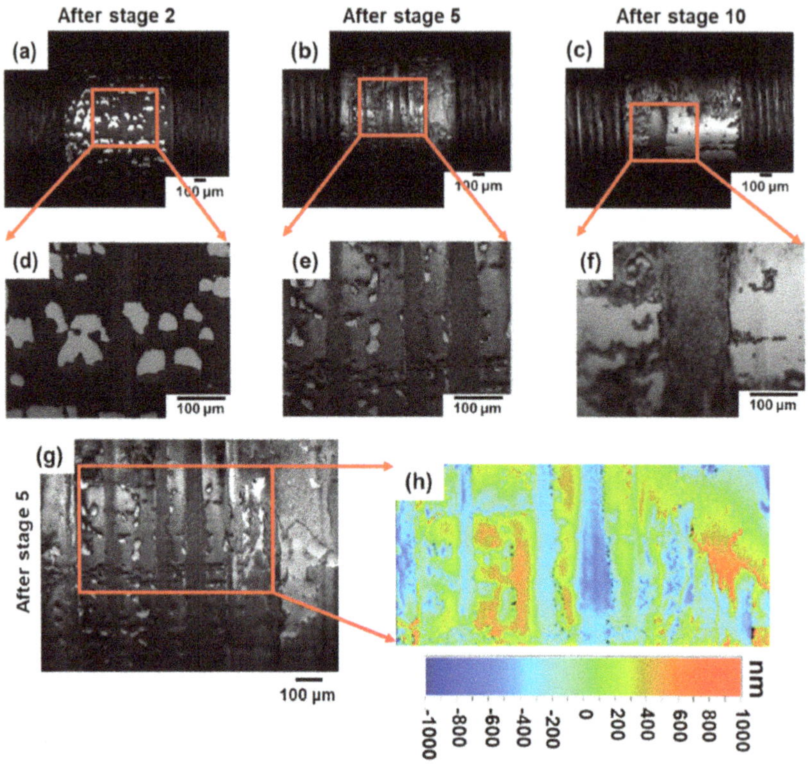

Figure 3.22. Evolution of the surface topography of roller lubricated by VHVI8+2% IL under low sliding (1.5% SRR) at low (a–c) and high (d–f) magnifications. (g) Morphology and (h) corresponding height map of the worn roller after test stage 5. Reprinted from [65], copyright (20210), with permission from Elsevier.

opposed to laboratory benchtop tests, will enhance customer confidence, increasing the industrialization of bio-based lubricants.

Much research has been conducted to investigate the tribological behavior of different nanoparticles and bio-based base stock. Nanoparticle morphology, concentration, and size are the critical parameters for enhancing the friction and wear resistance of additized lubricants. Detailed research on these parameters, and the optimum concentrations and suitable surfactants to formulate long-term stable lubricant solutions is necessary. Machine learning, simulation-based software, and newer preparation techniques should be incorporated to develop other reliable nano lubricants. Many cited studies focused on a few specific types of nanoparticle formulated lubricants and presented their tribological characteristics; however, the synergistic or antagonistic nature of multiple nanoparticles in the system and the impact on friction and wear mechanisms should be investigated. ILs have been promoted as the future of bio-lubricants by multiple researchers. The unique functionalities of these liquids, such as their nonflammable and nonvolatile natures, have attracted the attention of tribologists. Research performed to correlate molecular structure to tribological performance, hydrophobicity, and thermo-oxidative stability has been investigated; however, in-depth research is required in two specific directions: (1) create a method for the mass production of ILs, which will significantly reduce cost, and (2) develop novel, application-specific, base oil-soluble ILs. The toxicity and biodegradability of ILs should be investigated for environmental safety and security.

References

[1] Saidur R, Abdelaziz E A, Demirbas A, Hossain M S and Mekhilef S 2011 A review on biomass as a fuel for boilers *Renew. Sustain. Energy Rev.* **15** 2262–89

[2] Mang T and Dresel W 2007 *Lubricants and Lubrication* (Hoboken, NJ: Wiley)

[3] Schneider M P 2006 Plant-oil-based lubricants and hydraulic fluids *J. Sci. Food Agric.* **86** 1769–80

[4] Mahmoud A, Shuhaimi M and Abdel Samed M 2009 A combined process integration and fuel switching strategy for emissions reduction in chemical process plants *Energy* **34** 190–5

[5] Boroojeni K G, Amini M H, Nejadpak A, Iyengar S S, Hoseinzadeh B and Bak C L 2016 A theoretical bilevel control scheme for power networks with large-scale penetration of distributed renewable resources *IEEE Int. Conf. Electro Information Technology* 510–5

[6] Bart J C J, Gucciardi E and Cavallaro S 2013 Renewable lubricants *Biolubricants* **2013** 1–9

[7] Salimon J, Salih N and Yousif E 2010 Biolubricants: raw materials, chemical modifications and environmental benefits *Eur. J. Lipid Sci. Technol.* **112** 519–30

[8] Soni S and Agarwal M 2014 Lubricants from renewable energy sources—a review *Green Chem. Lett. Rev.* **7** 359–82

[9] Panchal T and Chauhan D 2014 Synthesis and characterization of bio lubricants from tobacco seed oil *Res. J. Agric. Environ. Manage.* **3** 97–105 http://apexjournal.org/rjaem/archive/2014/Feb/fulltext/Panchal et al.pdf

[10] Nagendramma P and Kaul S 2012 Development of ecofriendly/biodegradable lubricants: an overview *Renew. Sustain. Energy Rev.* **16** 764–74

[11] Aluyor E O, Obahiagbon K O and Ori-Jesu M 2009 Biodegradation of vegetable oils: a review *Sci. Res. Essays* **4** 543–48

[12] Cecilia J A, Plata D B, Saboya R M A, de Luna F M T, Cavalcante C L and Rodríguez-Castellón E 2020 An overview of the biolubricant production process: challenges and future perspectives *Processes* **8** 1–24

[13] Syahir A Z, Zulkifli N W M, Masjuki H H, Kalam M A, Alabdulkarem A, Gulzar M, Khuong L S and Harith M H 2017 A review on bio-based lubricants and their applications *J. Clean. Prod.* **168** 997–1016

[14] Lovell M, Higgs C F, Deshmukh P and Mobley A 2006 Increasing formability in sheet metal stamping operations using environmentally friendly lubricants *J. Mater. Process. Technol.* **177** 87–90

[15] Duzcukoglu H and Şahin Ö S 2011 Investigation of wear performance of canola oil containing boric acid under boundary friction condition *Tribol. Trans.* **54** 57–61

[16] Kržan B 2010 Study on the tribological performance of vegetable oils *Goriva Maz* **49** 360–7

[17] Bowden F P, Tabor D and Palmer F 1951 The friction and lubrication of solids *Am. J. Phys.* **19** 428

[18] Schober S and Mittelbach M 2004 The impact of antioxidants on biodiesel oxidation stability *Eur. J. Lipid Sci. Technol.* **106** 382–9

[19] Ruger C W, Klinker E J and Hammond E G 2002 Abilities of some antioxidants to stabilize soybean oil in industrial use conditions *J. Am. Oil Chem. Soc.* **79** 733–6

[20] Hewstone R K 1994 Environmental health aspects of lubricant additives *Sci. Total Environ.* **156** 243–54

[21] Kim B H, Mourhatch R and Aswath P B 2010 Properties of tribofilms formed with ashless dithiophosphate and zinc dialkyl dithiophosphate under extreme pressure conditions *Wear* **268** 579–91

[22] Feldmann D G and Hinrichs J 1997 Evaluation of the lubrication properties of biodegradable fluids and their potential to replace mineral oil in heavily loaded hydrostatic transmissions *ASTM Spec. Tech. Publ.* **1310** 220–9

[23] Reeves C J, Menezes P L, Jen T C and Lovell M R 2015 The influence of fatty acids on tribological and thermal properties of natural oils as sustainable biolubricants *Tribol. Int.* **90** 123–34

[24] Mannekote J K and Kailas S V 2012 The effect of oxidation on the tribological performance of few vegetable oils *J. Mater. Res. Technol.,* **1** 91–5

[25] Singh R K, Pandey S, Saxena R C, Thakre G D, Atray N and Ray S S 2015 Study of cystine schiff base esters as new environmentally benign multifunctional biolubricant additives *J. Ind. Eng. Chem.* **26** 149–56

[26] Gulzar M, Masjuki H, Varman M, Kalam M, Mufti R A, Zulkifli N, Yunus R and Zahid R 2015 Improving the AW/EP ability of chemically modified palm oil by adding CuO and MoS_2 nanoparticles *Tribol. Int.* **88** 271–9

[27] Reeves C J, Menezes P L, Lovell M R and Jen T C 2015 The influence of surface roughness and particulate size on the tribological performance of bio-based multi-functional hybrid lubricants *Tribol. Int.* **88** 40–55

[28] Lovell M R, Kabir M A, Menezes P L and Higgs C F 2010 Influence of boric acid additive size on green lubricant performance *Philos. Trans. R. Soc.* A **368** 4851–68

[29] Deshmukh P, Lovell M, Sawyer W G and Mobley A 2006 On the friction and wear performance of boric acid lubricant combinations in extended duration operations *Wear* **260** 1295–304

[30] Cermak S C, Bredsguard J W, John B L, McCalvin J S, Thompson T, Isbell K N, Feken K A, Isbell T A and Murray R E 2013 Synthesis and physical properties of new estolide esters *Ind. Crops Prod.* **46** 386–91

[31] Reeves C J, Menezes P L, Lovell M R and Jen T C 2013 The size effect of boron nitride particles on the tribological performance of biolubricants for energy conservation and sustainability *Tribol. Lett.* **51** 437–52

[32] Reeves C J, Siddaiah A and Menezes P L 2017 A review on the science and technology of natural and synthetic biolubricants *J. Bio-Tribo-Corros.* **3** 1–27

[33] Walden P 1914 Molecular weights and electrical conductivity of several fused salts *Bull. Acad. Imper. Sci. (St. Petersburg.)* **8** 405–22

[34] Chan B, Chang N and Grimmett M 1977 The synthesis and thermolysis of imidazole quaternary salts *Aust. J. Chem.* **30** 2005

[35] Chum H L, Koch V R, Miller L L and Osteryoung R A 1975 Electrochemical scrutiny of organometallic iron complexes and hexamethylbenzene in a room temperature molten salt *J. Am. Chem. Soc.* **97** 3264–5

[36] Wilkes J S, Levisky J A, Wilson R A and Hussey C L 1982 Dialkylimidazolium chloroaluminate melts: a new class of room-temperature ionic liquids for electrochemistry, spectroscopy and synthesis *Inorg. Chem.* **21** 1263–4

[37] Ye C, Liu W, Chen Y and Yu L 2001 Room-temperature ionic liquids: a novel versatile lubricant *Chem. Commun.* **21** 2244–5

[38] Somers A, Howlett P, MacFarlane D, Forsyth M, Somers A E, Howlett P C, MacFarlane D R and Forsyth M 2013 A review of ionic liquid lubricants *Lubricants* **1** 3–21

[39] Perkin S 2012 Ionic liquids in confined geometries *Phys. Chem. Chem. Phys.* **14** 5052

[40] Minami I 2009 Ionic liquids in tribology *Molecules* **14** 2286–305

[41] Palacio M and Bhushan B 2010 A review of ionic liquids for green molecular lubrication in nanotechnology *Tribol. Lett.* **40** 247–68

[42] Zhou Y and Qu J 2017 Ionic liquids as lubricant additives: a review *ACS Appl. Mater. Interfaces* **9** 3209–22

[43] Qu J *et al* 2012 Antiwear performance and mechanism of an oil-miscible ionic liquid as a lubricant additive *ACS Appl. Mater. Interfaces* **4** 997–1002

[44] Yu B, Bansal D G, Qu J, Sun X, Luo H, Dai S, Blau P J, Bunting B G, Mordukhovich G and Smolenski D J 2012 Oil-miscible and non-corrosive phosphonium-based ionic liquids as candidate lubricant additives *Wear* **289** 58–64

[45] Yao M, Fan M, Liang Y, Zhou F and Xia Y 2010 Imidazolium hexafluorophosphate ionic liquids as high temperature lubricants for steel-steel contacts *Wear* **268** 67–71

[46] Wang A, Chen L, Jiang D, Zeng H and Yan Z 2014 Vegetable oil-based ionic liquid microemulsion biolubricants: effect of integrated surfactants *Ind. Crops Prod.* **62** 515–21

[47] Amiri M and Khonsari M M 2010 On the thermodynamics of friction and wear—a review *Entropy* **12** 1021–49

[48] ASTM International 2011 ASTM D4172-21 Standard test method for wear preventive characteristics of lubricating fluid (four-ball method) *ASTM International*

[49] ASTM International 2003 ASTM D 2783-03 Standard test method for measurement of extreme-pressure properties of lubricating fluids (four-ball method) *ASTM International*

[50] Suresha B, Hemanth G, Rakesh A and Adarsh K M 2020 Tribological behaviour of neem oil with and without graphene nanoplatelets using four-ball tester *Adv. Tribol.* **2020** 1984931

[51] Baskar S, Sriram G and Arumugam S 2015 Experimental analysis on tribological behavior of nano based bio-lubricants using four ball tribometer *Tribol. Ind.* **37** 449–54

[52] Habibullah M, Masjuki H H, Kalam M A, Ashraful A M, Habib M A and Mobarak H M 2014 Effect of bio-lubricant on tribological characteristics of steel *Proc. Eng.* **90** 740–5

[53] Suresha B, Hemanth G, Rakesh A and Adarsh K M 2019 Tribological behaviour of pongamia oil as lubricant with and without halloysite nanotubes using four-ball tester *AIP Conf. Proc.* **2128** 030011

[54] Vithya P, Sriram G, Arumugam S, Sharma A and Venkata Sai Sathyanarayana Sastry S 2018 Experimental investigation on tribological characteristics of biodegradable refrigeration oil using pin on disc tribometer *IOP Conf. Ser.: Mater. Sci. Eng.* **390** 012052

[55] Singh Y, Pandey A K, Singh N K, Sharma A and Farooq A 2017 Sustainability of polanga oil based blended lubricant at different load and sliding distance *Energy Sources* A **39** 1970–7

[56] Singh Y, Singh P, Sharma A, Choudhary P, Singla A and Singh N K 2018 Optimization of wear and friction characteristics of *Phyllanthus emblica* seed oil based lubricant using response surface methodology *Egypt. J. Pet.* **27** 1145–55

[57] Cortes V, Sanchez K, Gonzalez R, Alcoutlabi M and Ortega J A 2020 The performance of SiO_2 and TiO_2 nanoparticles as lubricant additives in sunflower oil *Lubricants* **8** 10

[58] Akpan E I, Wetzel B and Friedrich K 2018 A fully biobased tribology material based on acrylic resin and short wood fibres *Tribol. Int.* **120** 381–90
[59] Xu Y, Hu X, Yuan K, Zhu G and Wang W 2014 Friction and wear behaviors of catalytic methylesterified bio-oil *Tribol. Int.* **71** 168–74
[60] Bahari A, Lewis R and Slatter T 2018 Friction and wear response of vegetable oils and their blends with mineral engine oil in a reciprocating sliding contact at severe contact conditions *Proc. Inst. Mech. Eng. J.* **232** 244–58
[61] Jason Y J J, How H G, Teoh Y H and Chuah H G 2020 A study on the tribological performance of nanolubricants *Processes* **8** 1–33
[62] Trzepieciński T 2020 Tribological performance of environmentally friendly bio-degradable lubricants based on a combination of boric acid and bio-based oils *Materials (Basel)* **13** 3892
[63] Talib N and Rahim E A 2016 The effect of tribology behavior on machining performances when using bio-based lubricant as a sustainable metalworking fluid *Proc. CIRP* **40** 504–8
[64] Roy S, Speed L, Viola M, Luo H, Leonard D and Qu J 2021 Oil miscible phosphonium-phosphate ionic liquid as novel antiwear and antipitting additive for low-viscosity rear axle lubricants *Wear* **466–467** 203588
[65] Roy S, Stump B C, Luo H, Leonard D and Qu J 2021 Why does a phosphonium-phosphinate ionic liquid protect the contact surfaces from wear and micropitting but increase vibration when used as an additive in rolling-sliding lubrication? *Tribol. Int.* **159** 106949
[66] Roy S, Ooi G T C and Sundararajan S 2018 Effect of retained austenite on micropitting behavior of carburized AISI 8620 steel under boundary lubrication *Materialia* **3** 192–201
[67] Cheenkachorn K 2013 A study of wear properties of different soybean oils *Energy Proc.* **42** 633–39
[68] Georgescu C, Solea L C and Deleanu L 2019 Additivation of vegetal oils for improving tribological characteristics *IOP Conf. Ser. Mater. Sci. Eng.* **514** 012012
[69] Habereder T, Moore D and Lang M 2009 Eco requirements for lubricant additives *Lubricant Additives: Chemistry and Applications* (Boca Raton, FL: CRC Press) pp 647–66
[70] Wu H, Zhao J, Xia W and Cheng X *et al* 2017 A study of the tribological behaviour of TiO_2 nano-additive water-based lubricants *Tribol. Int.* **109** 398–408
[71] Chang L, Zhang Z, Breidt C and Friedrich K 2005 Tribological properties of epoxy nanocomposites I. Enhancement of the wear resistance by nano-TiO_2 particles *Wear* **258** 141–8
[72] Jiao D, Zheng S, Wang Y, Guan R and Cao B 2011 The tribology properties of alumina/silica composite nanoparticles as lubricant additives *Appl. Surf. Sci.* **257** 5720–5
[73] Rapoport L, Leshchinsky V, Lvovsky M, Lapsker I, Volovik Y, Feldman Y, Popovitz-Biro R and Tenne R 2003 Superior tribological properties of powder materials with solid lubricant nanoparticles *Wear* **255** 794–800
[74] Chinas-Castillo F and Spikes H A 2003 Mechanism of action of colloidal solid dispersions *J. Tribol.* **125** 552–7
[75] Roy S, Jazaa Y and Sundararajan S 2019 Investigating the micropitting and wear performance of copper oxide and tungsten carbide nanofluids under boundary lubrication *Wear* **428–429** 55–63
[76] Kahlman L and Hutchings I M 1999 Effect of particulate contamination in grease-lubricated hybrid rolling bearings *Tribol. Trans.* **42** 842–50

[77] Xue Q, Liu W and Zhang Z 1997 Friction and wear properties of a surface-modified TiO_2 nanoparticle as an additive in liquid paraffin *Wear* **213** 29–32
[78] Zhang M, Wang X, Liu W and Fu X 2009 Performance and anti-wear mechanism of Cu nanoparticles as lubricating oil additives *Ind. Lubr. Tribol.* **61** 311–8
[79] Ji X, Chen Y, Zhao G, Wang X and Liu W 2011 Tribological properties of $CaCO_3$ nanoparticles as an additive in lithium grease *Tribol. Lett.* **41** 113–9
[80] Liu G, Li X, Qin B, Xing D, Guo Y and Fan R 2004 Investigation of the mending effect and mechanism of copper nano-particles on a tribologically stressed surface *Tribol. Lett.* **17** 961–6
[81] Tao X, Jiazheng Z and Kang X 1996 The ball-bearing effect of diamond nanoparticles as an oil additive *J. Phys. D: Appl. Phys.* **29** 2932–7
[82] Peng D X, Kang Y, Hwang R M, Shyr S S and Chang Y P 2009 Tribological properties of diamond and SiO_2 nanoparticles added in paraffin *Tribol. Int.* **42** 911–7
[83] Berman D, Erdemir A, Zinovev A V and Sumant A V 2015 Nanoscale friction properties of graphene and graphene oxide *Diam. Relat. Mater.* **54** 91–6
[84] Su Y, Gong L and Chen D 2015 An investigation on tribological properties and lubrication mechanism of graphite nanoparticles as vegetable based oil additive *J. Nanomater.* **2015** 276753
[85] Lee J, Cho S, Hwang Y, Lee C and Kim S H 2007 Enhancement of lubrication properties of nano-oil by controlling the amount of fullerene nanoparticle additives *Tribol. Lett.* **28** 203–8
[86] Wang R, Zhang Y and Wu H 2021 Tribological properties of onion like fullerenes and $NiFe_2O_4$ nanocomposites in reciprocating motion *Front. Energy* **15** 208–12
[87] Cristea G C, Dima C, Georgescu C, Dima D, Deleanu L and Solea L C 2018 Evaluating lubrication capability of soybean oil with nano carbon additive *Tribol. Ind.* **40** 66–72
[88] Zhang G, Xu Y, Xiang X, Zheng G, Zeng X, Li Z, Ren T and Zhang Y 2018 Tribological performances of highly dispersed graphene oxide derivatives in vegetable oil *Tribol. Int.* **126** 39–48
[89] Roy S and Ghosh A 2013 High speed turning of AISI 4140 steel using nanofluid through twin jet SQL system *ASME 2013 Int. Manufacturing Science and Engineering Conf. Collocated with the 41st North American Manufacturing Research Conf., MSEC 2013* vol 2
[90] Roy S and Ghosh A 2014 High-speed turning of AISI 4140 steel by multi layered TiN top-coated insert with minimum quantity lubrication technology and assessment of near tool-tip temperature using infrared thermography *Proc. Inst. Mech. Eng.* B **228** 1058–67
[91] Pan Q and Zhang X 2010 Synthesis and tribological behavior of oil-soluble Cu nanoparticles as additive in SF15W/40 lubricating oil *Rare Met. Mater. Eng.* **39** 1711–4
[92] Ma J, Mo Y and Bai M 2009 Effect of Ag nanoparticles additive on the tribological behavior of multialkylated cyclopentanes (MACs) *Wear* **266** 627–31
[93] Padgurskas J, Rukuiža R, Kreivaitis R, Asadauskas S J and Bražinskiene D 2009 Tribologic behaviour and suspension stability of iron and copper nanoparticles in rapeseed and mineral oils *Tribol.—Mater. Surfaces Interfaces* **3** 97–102
[94] Padgurskas J, Rukuiza R, Prosyčevas I and Kreivaitis R 2013 Tribological properties of lubricant additives of Fe, Cu and Co nanoparticles *Tribol. Int.* **60** 224–32
[95] Zareh-Desari B and Davoodi B 2016 Assessing the lubrication performance of vegetable oil-based nano-lubricants for environmentally conscious metal forming processes *J. Clean. Prod.* **135** 1198–209

[96] Rajaganapathy C, Vasudevan D and Murugapoopathi S 2020 Tribological and rheological properties of palm and brassica oil with inclusion of CuO and TiO_2 additives *Mater. Today Proc.* **37** 207–13
[97] Peña-Parás L, Taha-Tijerina J, Garza L, Maldonado-Cortés D, Michalczewski R and Lapray C 2015 Effect of CuO and Al_2O_3 nanoparticle additives on the tribological behavior of fully formulated oils *Wear* **332–333** 1256–61
[98] Cristea G C, Cazamir D, Dima D, Georgescu C and Deleanu L 2018 Influence of TiO_2 as nano additive in rapeseed oil *IOP Conf. Ser.: Mater. Sci. Eng.* **444** 022011
[99] Ilie F and Covaliu C 2016 Tribological properties of the lubricant containing titanium dioxide nanoparticles as an additive *Lubricants* **4** 12
[100] Trajano M F, Moura E I F, Ribeiro K S B and Alves S M 2014 Study of oxide nanoparticles as additives for vegetable lubricants *Mater. Res.* **17** 1124–8
[101] Ma S, Zheng S, Cao D and Guo H 2010 Anti-wear and friction performance of ZrO_2 nanoparticles as lubricant additive *Particuology* **8** 468–72
[102] Xie H, Jiang B, Hu X, Peng C, Guo H and Pan F 2017 Synergistic effect of MoS_2 and SiO_2 nanoparticles as lubricant additives for magnesium alloy–steel contacts *Nanomaterials* **7** 154
[103] Aldana P U, Dassenoy F, Vacher B, Le Mogne T and Thiebaut B 2016 WS_2 nanoparticles anti-wear and friction reducing properties on rough surfaces in the presence of ZDDP additive *Tribol. Int.* **102** 213–21
[104] Abdullah M I H C, Bin Abdollah M F, Amiruddin H, Tamaldin N and Nuri N R M 2017 Tribological effects of nano-based engine oil diluted with biodiesel fuel *Int. J. Surf. Sci. Eng.* **11** 12–22
[105] Shi S C 2016 Tribological performance of green lubricant enhanced by sulfidation IF-MoS_2 *Materials (Basel)* **9** 856
[106] Abdullah M I H C, Bin Abdollah M F, Amiruddin H, Tamaldin N and Nuri N R M 2013 Optimization of tribological performance of hBN/Al_2O_3 nanoparticles as engine oil additives *Proc. Eng.* **68** 313–9
[107] He X, Xiao H, Choi H, Díaz A, Mosby B, Clearfield A and Liang H 2014 α-Zirconium phosphate nanoplatelets as lubricant additives *Colloids Surf.* A **452** 32–8
[108] Sen Zhang B, Xu B S, Xu Y, Gao F, Shi P J and Wu Y X 2011 CU nanoparticles effect on the tribological properties of hydrosilicate powders as lubricant additive for steelsteel contacts *Tribol. Int.* **44** 878–86
[109] Gu K C, Chen B S, Wang X M, Wang J, Fang J H, Wu J and Huang L C 2014 Preparation, friction and wear properties of hydrophobic lanthanum borate nanorods in rapeseed oil *Trans. Nonferrous Met. Soc. China (Engl. Ed.)* **24** 3578–84
[110] Zhao F, Bai Z, Fu Y, Zhao D and Yan C 2012 Tribological properties of serpentine, La $(OH)_3$ and their composite particles as lubricant additives *Wear* **288** 72–7
[111] ji X, Chen Y, Wang X and Liu W 2012 Tribological behaviors of novel tri(hydroxymethyl) propane esters containing boron and nitrogen as lubricant additives in rapeseed oil *Ind. Lubr. Tribol.* **64** 315–20
[112] Dubey M K, Bijwe J and Ramakumar S S V 2015 Nano-PTFE: new entrant as a very promising EP additive *Tribol. Int.* **87** 121–31
[113] Sen Zhang B, Xu B S, Xu Y, Gao F, Shi P J and Wu Y X 2011 CU nanoparticles effect on the tribological properties of hydrosilicate powders as lubricant additive for steel–steel contacts *Tribol. Int.* **44** 878–86

[114] Qian J, Zhang Y, Wang L and Xing J 2011 Study on lubrication properties of modified nano ZnO in base oil *China Pet. Process. Petrochem. Technol.* **13** 69–73
[115] Song X, Zheng S, Zhang J, Li W, Chen Q and Cao B 2012 Synthesis of monodispersed $ZnAl_2O_4$ nanoparticles and their tribology properties as lubricant additives *Mater. Res. Bull.* **47** 4305–10
[116] Cristea G C, Georgescu C, Dima D and Deleanu L 2018 Influence of graphene as additive in soybean oil *IOP Conf. Ser.: Mater. Sci. Eng.* **444** 022012
[117] Hernández Battez A, González R, Felgueroso D, Fernández J E, del Rocío Fernández M, García M A and Peñuelas I 2007 Wear prevention behaviour of nanoparticle suspension under extreme pressure conditions *Wear* **263** 1568–74
[118] Hernandez Battez A, Fernandez Rico J E, Navas Arias A, Viesca Rodriguez J L, Chou Rodriguez R and Diaz Fernandez J M 2006 The tribological behaviour of ZnO nanoparticles as an additive to PAO6 *Wear* **261** 256–63
[119] Amiril S A S, Rahim E A and Syahrullail S 2017 A review on ionic liquids as sustainable lubricants in manufacturing and engineering: recent research, performance, and applications *J. Clean. Prod.* **168** 1571–89
[120] Espinosa T, Sanes J and Bermúdez M D 2016 New alkylether-thiazolium room-temperature ionic liquid lubricants: surface interactions and tribological performance *ACS Appl. Mater. Interfaces* **8** 18631–9
[121] Chiappe C and Pomelli C S 2016 Ionic liquids and green chemistry *Analytical Applications of Ionic Liquids* ed M Koel (Singapore: World Scientific) pp 385–404
[122] Blanco D, González R, Viesca J L, Fernández-González A, Bartolomé M and Hernández Battez A 2017 Antifriction and antiwear properties of an ionic liquid with fluorine-containing anion used as lubricant additive *Tribol. Lett.* **65** 66
[123] Schmidt T A, Gastelum N S, Nguyen Q T, Schumacher B L and Sah R L 2007 Boundary lubrication of articular cartilage: role of synovial fluid constituents *Arthritis Rheumatol.* **56** 882–91
[124] Lei Z, Chen B, Koo Y M and Macfarlane D R 2017 Introduction: ionic liquids *Chem. Rev.* **117** 6633–5
[125] Liu W, Ye C, Gong Q, Wang H and Wang P 2002 Tribological performance of room-temperature ionic liquids as lubricant *Tribol. Lett.* **13** 81–5
[126] Chen Y, Ye C, Wang H and Liu W 2003 Tribological performance of an ionic liquid as a lubricant for steel/aluminum contacts *J. Synth. Lubr.* **20** 217–25
[127] Mu Z, Zhou F, Zhang S, Liang Y and Liu W 2005 Effect of the functional groups in ionic liquid molecules on the friction and wear behavior of aluminum alloy in lubricated aluminum-on-steel contact *Tribol. Int.* **38** 725–31
[128] Jiménez A E, Bermúdez M D, Carrión F J and Martínez-Nicolás G 2006 Room temperature ionic liquids as lubricant additives in steel–aluminium contacts: influence of sliding velocity, normal load and temperature *Wear* **261** 347–59
[129] Kamimura H, Kubo T, Minami I and Mori S 2007 Effect and mechanism of additives for ionic liquids as new lubricants *Tribol. Int.* **40** 620–5
[130] Qu J, Blau P J, Dai S, Luo H M, Meyer H M and Truhan J J 2009 Tribological characteristics of aluminum alloys sliding against steel lubricated by ammonium and imidazolium ionic liquids *Wear* **267** 1226–31
[131] Murakami T, Kaneda K, Nakano M, Korenaga A, Mano H and Sasaki S 2008 Tribological properties of Fe_7Mo_6-based alloy under two ionic liquid lubrications *Tribol. Int.* **41** 1083–9

[132] Iglesias P, Bermúdez M D, Carrión F J and Martínez-Nicolás G 2004 Friction and wear of aluminium–steel contacts lubricated with ordered fluids-neutral and ionic liquid crystals as oil additives *Wear* **256** 386–92

[133] Bermúdez M D and Jiménez A E 2007 Surface interactions and tribochemical processes in ionic liquid lubrication of aluminium-steel contacts *Int. J. Surf. Sci. Eng.* **1** 100–10

[134] Jiménez A E and Bermúdez M D 2012 Short alkyl chain imidazolium ionic liquid additives in lubrication of three aluminium alloys with synthetic ester oil *Tribol.—Mater. Surf. Interface.* **6** 109–15

[135] Mu Z, Wang X, Zhang S, Liang Y, Bao M and Liu W 2008 Investigation of tribological behavior of Al–Si alloy against steel lubricated with ionic liquids of 1-diethylphosphonyl-*n*-propyi-3-alkylimidazolium tetrafluoroborate *J. Tribol.* **130** 034501

[136] Mu Z, Liu W, Zhang S and Zhou F 2004 Functional room-temperature ionic liquids as lubricants for an aluminum-on-steel system *Chem. Lett.* **33** 524–5

[137] Qu J, Truhan J J, Dai S, Luo H and Blau P J 2006 Ionic liquids with ammonium cations as lubricants or additives *Tribol. Lett.* **22** 207–14

[138] Zeng Z, Chen Y, Wang D and Zhang J 2007 Tribological behaviors of amorphous Cr coatings electrodeposited from Cr(III) baths under ionic liquid lubrication *Electrochem. Solid-State Lett.* **10** D85

[139] Liu X, Zhou F, Liang Y and Liu W 2006 Benzotriazole as the additive for ionic liquid lubricant: one pathway towards actual application of ionic liquids *Tribol. Lett.* **23** 191–6

[140] Xia Y, Wang L, Liu X and Qiao Y 2008 Tribological properties of phosphor bronze and nanocrystalline nickel coatings under PAO + MoDTC and ionic liquid lubricated condition *Tribol. Lett.* **31** 149–58

[141] Xia Y, Sasaki S, Murakami T, Nakano M, Shi L and Wang H 2007 Ionic liquid lubrication of electrodeposited nickel–Si_3N_4 composite coatings *Wear* **262** 765–71

[142] Xia Y, Wang L, Liu X and Qiao Y 2008 A comparative study on the tribological behavior of nanocrystalline nickel and coarse-grained nickel coatings under ionic liquid lubrication *Tribol. Lett.* **30** 151–7

[143] Sanes J, Carrión F J, Bermúdez M D and Martínez-Nicolás G 2006 Ionic liquids as lubricants of polystyrene and polyamide 6-steel contacts. Preparation and properties of new polymer-ionic liquid dispersions *Tribol. Lett.* **21** 121–33

[144] Sanes J, Carrión-Vilches F J and Bermúdez M D 2007 New epoxy-ionic liquid dispersions. Room temperature ionic liquid as lubricant of epoxy resin–stainless steel contacts *E-Polymers* **2007** 005

[145] Carrión F J, Sanes J and Bermúdez M D 2007 Effect of ionic liquid on the structure and tribological properties of polycarbonate-zinc oxide nanodispersion *Mater. Lett.* **61** 4531–5

[146] Kondo H 2008 Protic ionic liquids with ammonium salts as lubricants for magnetic thin film media *Tribol. Lett.* **31** 211–8

[147] Torimoto T, Tsuda T, Okazaki K I and Kuwabata S 2010 New frontiers in materials science opened by ionic liquids *Adv. Mater.* **22** 1196–221

[148] García A, González R, Hernández Battez A, Viesca J L, Monge R, Fernández-González A and Hadfield M 2014 Ionic liquids as a neat lubricant applied to steel–steel contacts *Tribol. Int.* **72** 42–50

[149] Freire M G, Neves C M S S, Marrucho I M, Coutinho J A P and Fernandes A M 2010 Hydrolysis of tetrafluoroborate and hexafluorophosphate counter ions in imidazolium-based ionic liquids *J. Phys. Chem.* A **114** 3744–9

[150] Uerdingen M, Treber C, Balser M, Schmitt G and Werner C 2005 Corrosion behaviour of ionic liquids *Green Chem.* **7** 321–5

[151] Nosonovsky M and Bhushan B 2010 Green tribology: principles, research areas and challenges *Philos. Trans. R. Soc.* A **368** 4677–94

[152] Jiménez A E, Bermúdez M D, Iglesias P, Carrión F J and Martínez-Nicolás G 2006 1-*N*-alkyl-3-methylimidazolium ionic liquids as neat lubricants and lubricant additives in steel-aluminium contacts *Wear* **260** 766–82

[153] González R, Hernández Battez A, Blanco D, Viesca J L and Fernández-González A 2010 Lubrication of TiN, CrN and DLC PVD coatings with 1-butyl-1-methylpyrrolidinium tris (pentafluoroethyl)trifluorophosphate *Tribol. Lett.* **40** 269–77

[154] Blanco D, Battez A H, Viesca J L, González R and Fernández-González A 2011 Lubrication of CrN coating with ethyl-dimethyl-2-methoxyethylammonium tris(pentafluoroethyl)trifluorophosphate ionic liquid as additive to PAO 6 *Tribol. Lett.* **41** 295–302

[155] Viesca J L, García A, Hernández Battez A, González R, Monge R, Fernández-González A and Hadfield M 2013 FAP-anion ionic liquids used in the lubrication of a steel–steel contact *Tribol. Lett.* **52** 431–7

[156] Minami I, Kita M, Kubo T, Nanao H and Mori S 2008 The tribological properties of ionic liquids composed of trifluorotris(pentafluoroethyl) phosphate as a hydrophobic anion *Tribol. Lett.* **30** 215–23

[157] Minami I, Inada T, Sasaki R and Nanao H 2010 Tribo-chemistry of phosphonium-derived ionic liquids *Tribol. Lett.* **40** 225–35

[158] Somers A E, Khemchandani B, Howlett P C, Sun J, MacFarlane D R and Forsyth M 2013 Ionic liquids as antiwear additives in base oils: influence of structure on miscibility and antiwear performance for steel on aluminum *ACS Appl. Mater. Interfaces* **5** 11544–53

[159] Smiglak M, Metlen A and Rogers R D 2007 The second evolution of ionic liquids: from solvents and separations to advanced materials—energetic examples from the ionic liquid cookbook *ACC. Chem. Res.* **40** 1182–92

[160] Costa S P F, Azevedo A M O, Pinto P C A G and Saraiva M L M F S 2017 Environmental impact of ionic liquids: recent advances in (eco)toxicology and (bio)degradability *ChemSusChem* **10** 2321–47

[161] Jiang C, Li W, Nian J, Lou W and Wang X 2018 Tribological evaluation of environmentally friendly ionic liquids derived from renewable biomaterials *Friction* **6** 208–18

[162] Mustahil N A, Baharuddin S H, Abdullah A A, Reddy A V B, Abdul Mutalib M I and Moniruzzaman M 2019 Synthesis, characterization, ecotoxicity and biodegradability evaluations of novel biocompatible surface active lauroyl sarcosinate ionic liquids *Chemosphere* **229** 349–57

[163] Reeves C J, Garvey S, Menezes P L, Dietz M, Jen T C and Lovell M R 2012 Tribological performance of environmentally friendly ionic liquid lubricants *Proc. of the ASME/STLE 2012 Int. Joint Tribology Conf. ASME/STLE 2012 Int. Joint Tribology Conf.* 355–7

[164] Parmentier D, Metz S J and Kroon M C 2013 Tetraalkylammonium oleate and linoleate based ionic liquids: promising extractants for metal salts *Green Chem.* **15** 205–9

[165] Gusain R, Dhingra S and Khatri O P 2016 Fatty-acid-constituted halogen-free ionic liquids as renewable, environmentally friendly, and high-performance lubricant additives *Ind. Eng. Chem. Res.* **55** 856–65

[166] Faes J, González R, Battez A H, Blanco D, Fernández-González A and Viesca J L 2021 Friction, wear and corrosion behavior of environmentally-friendly fatty acid ionic liquids *Coatings* **11** 1–16

[167] Gusain R and Khatri O P 2016 Fatty acid ionic liquids as environmentally friendly lubricants for low friction and wear *RSC Adv.* **6** 3462–9

[168] Khatri P K, Aathira M S, Thakre G D and Jain S L 2018 Synthesis and tribological behavior of fatty acid constituted tetramethylguanidinium (TMG) ionic liquids for a steel/steel contact *Mater. Sci. Eng.* C **91** 208–17

[169] Blanco D, Rivera N, Oulego P, Díaz M, González R and Battez A H 2019 Novel fatty acid anion-based ionic liquids: contact angle, surface tension, polarity fraction and spreading parameter *J. Mol. Liq.* **288** 110995

[170] Rivera N, Blanco D, Viesca J L, Fernández-González A, González R and Battez A H 2019 Tribological performance of three fatty acid anion-based ionic liquids (FAILs) used as lubricant additive *J. Mol. Liq.* **296** 111881

[171] Khan A, Gusain R, Sahai M and Khatri O P 2019 Fatty acids-derived protic ionic liquids as lubricant additive to synthetic lube base oil for enhancement of tribological properties *J. Mol. Liq.* **293** 111444

[172] Rivera N, García A, Fernández-González A, Blanco D, González R and Battez A H 2019 Tribological behavior of three fatty acid ionic liquids in the lubrication of different material pairs *J. Mol. Liq.* **296** 111858

[173] Gusain R, Khan A and Khatri O P 2020 Fatty acid-derived ionic liquids as renewable lubricant additives: effect of chain length and unsaturation *J. Mol. Liq.* **301** 112322

[174] Sernaglia M, Blanco D, Battez A H, Viesca J L, González R and Bartolomé M 2020 Two fatty acid anion-based ionic liquids—part I: physicochemical properties and tribological behavior as neat lubricants *J. Mol. Liq.* **305** 112827

[175] Reeves C J, Siddaiah A and Menezes P L 2017 Ionic liquids: a plausible future of bio-lubricants *J. Bio-Tribo-Corros.* **3** 18

[176] Hou X D, Liu Q P, Smith T J, Li N and Zong M H 2013 Evaluation of toxicity and biodegradability of cholinium amino acids ionic liquids *PLoS One* **8** e59145

[177] Stump B C, Zhou Y, Luo H M, Leonard D N, Viola M B and Qu J 2019 New functionality of ionic liquids as lubricant additives: mitigating rolling contact fatigue *ACS Appl. Mater. Interfaces* **11** 30484–92

Chapter 4

Multiphase bio-lubricants for engine components

Dipen Kumar Rajak, D D Pagar and Pradeep L Menezes

Internal combustion engines (ICEs) have always been accountable for mechanical power generation in most transportation applications such as automobile, marine, and lightweight aviation industries. Friction between mechanical components of ICEs accounts for a high proportion of power losses. The study of tribological surfaces in relative contact has revealed various approaches to improve efficiency. Introduction of certain nanoparticle additives in lubricants has reduced frictional losses. In this chapter, the application of bio-based lubricants for engine components such as gears, cams, tappets, and piston rings is presented.

4.1 History

Tribology first entered the literature in the mid-1960s, explaining the occurrence of friction, wear, and lubrication. Procured from the Greek 'tribos', it is interpreted as 'the act of rubbing', while the suffix 'ology' refers to the study or science or knowledge of a certain subject. Thus, the word tribology signifies 'the science of rubbing' [1]. Looking back into the Palaeolithic period, humans invented fire by means of friction, which has been considered one of the greatest discoveries of all time. Thereafter, when humans started traveling around the globe to explore places, the discovery of wheels took place to truncate frictional losses in translational motion. They worked out that there was much less friction in rolling motion compared to sliding. This signifies that the phenomenon of friction was perceived by humans long ago. However, it was first hypothesized by Leonardo da Vinci proposing the term coefficient of friction as proportionality between frictional force and normal force far before Newton's disclosure of the concept of force. Later on, Guillaume Amontons in 1699 and Charles August Coulomb in 1785 put forward the laws of friction, which were later explained by F Philip Bowden and David Tabor in 1950 [2, 3].

doi:10.1088/978-0-7503-3435-8ch4

Nevertheless, a prolonged investigation into the wear phenomenon was initiated in Germany during the 1930s. Ultimately, the International Research Group on Wear of Engineering Materials, formed by the Organisation for Economic Co-operation and Developement (OECD) countries in 1969, united researchers around the globe to propose standards and regulations to formulate wear. In 1980, the American Society of Mechanical Engineering (ASME) published the *Wear Control Handbook* for generalizing mechanical design considerations [4, 5]. It was believed that for constructing gigantic structures, ancient Egyptians first made use of water as a lubricant to slide massive rocks over the ground dragged by hundreds of people. Later on, vegetable oils and animal fats were introduced as lubricants, which were then replaced by more efficient petroleum-based oils discovered in the 20th century. Presently, various lubricants considering their formulated viscosity index (VI) for applications have evolved [6–8].

4.2 Introduction

In a working environment where solid surfaces interact with each other in relative motion, they degrade mechanically as a consequence of friction, which also explains the occurrence of wear. Therefore, at such an event, lubrication is introduced where a fluid film acts as a medium between two interacting surfaces to reduce the surface contact. The study of friction, wear, and lubrication altogether describes the multidisciplinary phenomenon of tribology [9].

4.2.1 Friction, wear and lubrication for tribological surfaces

All engineered surfaces are rough, and the roughness appears in the form of protuberances when observed at microscopic levels, which are generally termed as asperities. Fundamentally, roughness of the material is a variation in the amplitude of these asperities with respect to a reference plane. The manifestation of surface roughness is due to several different machining processes and working environments encountered by the material during fabrication [10]. Every solid surface is featured by its roughness as variation in asperities. This asperity height ranges from a few nanometers to several micrometers. The surface topography of a material comprehends the study of variation in the amplitude of the peaks and valleys of such asperities, their spacing, and orientation. It is a property possessed by a material that is considered to be acquired during its manufacturing. For instance, the outermost surface of a strain-hardened layer of a machined component is contained with impurities. These impurities include dust particles, remnants from atmospheric oxidation, cutting tools, and their lubrication [11, 12]. One of the conventional procedures to measure the surface roughness was using a diamond stylus profilometer. The diameter of a conical-shaped stylus ranges from several nanometers to a few micrometers. The stylus is subjected to run transversely over the surface of a material at a controlled speed. The profile generated by the vertical displacement of a stylus is recorded to give the measure of surface roughness as an output. This method incorporated direct contact with the material surface, which sometimes tends to disrupt the surface topography of some materials, besides the process being lengthy [13]. Some modern equipment, such as the atomic force microscope (AFM), utilizes advanced scanning technology to produce high-resolution 3-D images of surfaces. Also, there are non-contacting type instruments such as optical interferometers that use white light or laser beam technology to measure surface roughness [14, 15].

Surface irregularities correspond to surface roughness, consequently developing friction when rubbed on their counterparts. Tribological surfaces tend to wear away with a deleterious impact of frictional forces. Therefore, these surfaces are treated with a layer of lubricants to minimize friction and surface wear. The layer of lubricants can be liquid and/or solid lubricants. For example, the multilayered coating of tungsten carbide reinforced with diamond-like carbon (WC/a-C) is employed for steel and silicon substrates. The formation of a significant layer of graphitized carbon and WO_3 results in lowering the friction coefficient to 0.05 as well as enhancing anti-wear properties [16]. A significant reduction in friction coefficient by 56% was observed with the presence of nano-sized hexagonal boron nitride (h-BN) powder and graphene in the nickel matrix. The development of the graphene layer on a worn out surface improved load-bearing capacity and anti-wear properties [17]. A bearing steel surface was modified by distributing molybdenum disulfide (MoS_2) and graphite particles to form a layer. It enhances bearing strength with an increase in micro-hardness by 20% and reduces friction factor [18]. The excellent wear-resistant surface was produced by dispersing nanoparticles of copper flakes. Copper nanoparticles possess remarkable superplasticity, self-healing, and anti-friction characteristics to protect a surface from wear at high speeds and intense loading conditions [19].

4.2.2 Friction

The phenomenon of friction is simply defined as resistance to motion. Here, motion is considered as relative motion between two objects moving tangentially over one another, maintaining close contact. Friction is not a property possessed by any material; rather, it is an incidence of energy loss as a responsive action against the relative motion [1].

The coefficient of friction depends on the type of materials and surface topography. The coefficient of friction is expressed as the ratio between frictional force and the force normal to the contacting surfaces,

$$\mu = \frac{F}{N} \tag{4.1}$$

where μ is the coefficient of friction, F is friction force, and N is the normal force. The greater the value of friction coefficient between the contacting surfaces, the larger will be the force required to move the body along that surface. The value of the friction coefficient for the material is independent of the apparent area of contact, however, it is minimized by employing various surface treatments or by introducing a lubrication film at the interface [20–23].

The friction phenomenon is further explained with the help of micro-asperities present on the surface of any material when inspected at the micro-scale, as shown in figure 4.1. When two surfaces meet, they make a conformal contact at some of these highest points, randomly shaped asperities that are subjected to carry the entire load. Ascribed to a normal force, these micro-asperities develop the highest amount of local contact stresses at the point of contact. These interlocked asperities hold the surfaces together, hindering the movement of bodies in the direction of applied shear force. In such an instance, the shearing force, on account of the applied force, has to be greater than the normal force to move these glued surfaces [24, 25]. The energy required for the application of this force is sometimes translated into heat or plastically deforms the micro-asperities at the contacting surfaces. Therefore, sometimes friction is considered a detrimental phenomenon as it further wears out the material surface.

However, in figure 4.1(b), the course of the event of interlocking asperities restrains block A from descending and helps to maintain its state of rest. Conveying this feature, friction is beneficial to keep objects at rest in myriad fields of applications. However, for a moving object, such as a vehicle, it aids the motion by sticking to the road surface based on all the parameters from equation (4.1). In the absence of friction, automobiles would fail to function on road surfaces since the traction between the tires and the road surface keeps the wheels rotating. Moreover, in brake assembly, friction between the brake pads and rotors used decelerates the vehicle while braking [26]. Internal friction exhibited by solid materials resists the flow of atoms inside the material. Even in fluids, internal friction exists, and generally, it is termed as a viscosity where viscous force resists the motion of fluid layers in various directions. More viscous fluid signifies greater resistance offered to shear action between the fluid layers [27].

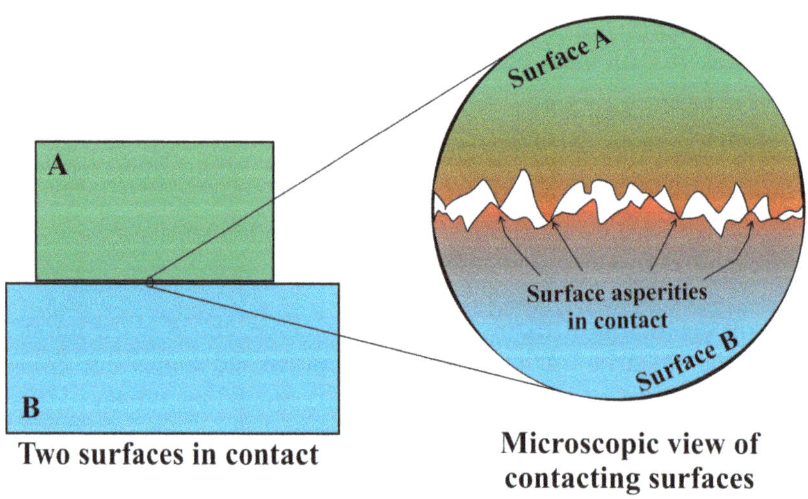

Figure 4.1. Schematic of surface interaction and surface asperities.

4.2.3 Wear

Wear is a progressive loss of material from its precursory dimension, position, and physical structure on account of friction. It is one of the consequences of plastic deformation that occurs between asperities of contacting surfaces in relative motion [1]. According to the way it affects the surface topography of the material, the wear phenomenon is further explained in the following categories.

4.2.3.1 Adhesive wear

On account of normal force, the molecules present at the closely contacting surfaces interact with each other creating a bond at the interface. Further rolling or sliding between two surfaces plastically deforms the material by shearing off this bond. This action transfers the layer of material with smaller shear strength to the counter-surface. Galling, mostly observed in sliding contact bearings, is considered as a type of adhesive wear [28].

4.2.3.2 Abrasive wear

Under the influence of both compressive and shear stresses, asperities of material with greater hardness cause a ploughing effect on the surface of softer material. Sometimes, contaminations in lubrication such as soot, dust particles, or broken asperities of harder surface also damage the protective layer of softer counter-surface and plastically deform it [29].

4.2.3.3 Fatigue wear

Cyclic loading causes stress concentration in certain contacting regions. Micro-structural deformities in the material further develop failure phenomena such as crystallographic dislocation, delamination, and micro-cracks generation [9].

4.2.3.4 Corrosive wear

Material degradation based on chemical or electrochemical interaction together with mechanical rubbing between contacting surfaces leads to tribological corrosion. Surface erosion attributed to wear is more vulnerable to corrosion as it loses its protective layer and is exposed to oxidation. Moreover, a corroded surface forms oxide debris at the interface of contacting surfaces in relative motion. The oscillatory motion of small amplitude causes surface erosion because of abrasive wear, which is termed fretting corrosion [2, 30, 31].

4.2.4 Lubrication

In an ICE, high stresses and wear in mechanical components develop noises and vibrations while operating. Therefore, a medium is introduced in the form of an emulsion, powder, contaminants, oils, or various types of fluids (including air) between the contacting surfaces in relative motion [32]. A lubricant is a low shear strength material that resides in the gap between the asperities of contacting surfaces. It is deployed to minimize friction and prevent rubbing surfaces from corrosion. It fosters the successful transfer of dynamic or static loads, control over friction coefficient, heat dissipation, and energy conservation. Furthermore, an ideal lubricant has to withstand higher temperatures without affecting its viscosity and must contain oxidation inhibitors [20, 21].

The effectiveness of a lubricant is a measure of its capacity to produce a layer of a certain thickness which is dependent upon its physical and chemical properties. Therefore, while selecting a lubricant its compatibility with the working temperature, loading conditions, presence of moisture, and type of relative motion between the mechanical components (sliding or rolling) have to be taken into the consideration. Base oils are the largest constituents of a lubricant, typically from 70 to 95%; accordingly they are classified as shown in figure 4.2 [33, 34].

Lubricants can be classified based on their physical state as solid, liquid or quasi-solid. Liquid lubricants have widespread domestic and industrial applications. Quasi-solid lubricants are needed to retain their phase at the desired position for a significant amount of time. These include greases, waxes, and compounds of oil and fats utilized for high viscosity, high rolling friction applications. Greases and liquid lubricants are extensively employed for long run applications. However, solid lubricants are employed where greases and liquid lubricants fail, such as micro-gravity, high vacuum, extreme temperature, and pressure conditions. MoS_2, graphite, and boric acid are the most commonly used solid lubricants. Solid lubricants are sometimes applied as additives in grease and liquid lubricants [32].

4.2.4.1 Lubrication regimes

Depending on the lubricant film thickness (h) and frictional coefficient (μ) between two contacting surfaces, a modified Stribeck's curve presented in figure 4.3 differentiates three lubrication regimes, namely boundary lubrication (BL), mixed lubrication (ML), and hydrodynamic lubrication (HL). This curve estimates variation in frictional coefficient (μ) and lubricant film thickness (h) as a function of Hersey number ($\eta\frac{N}{P}$), where η is lubricant viscosity, N is relative speed, P is normal applied load [14, 35, 36].

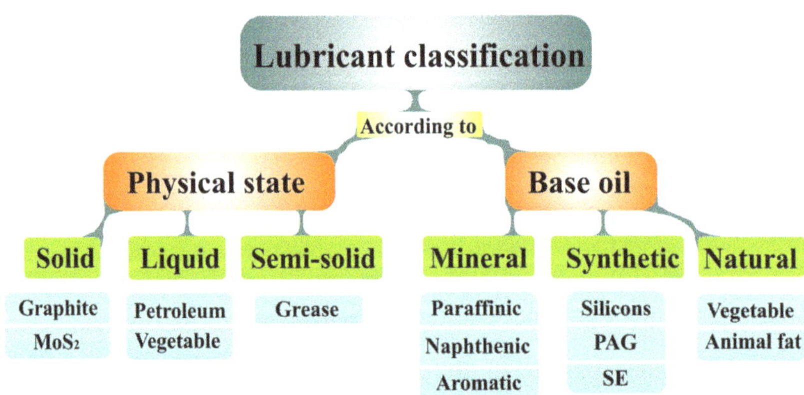

Figure 4.2. Classification of lubricants.

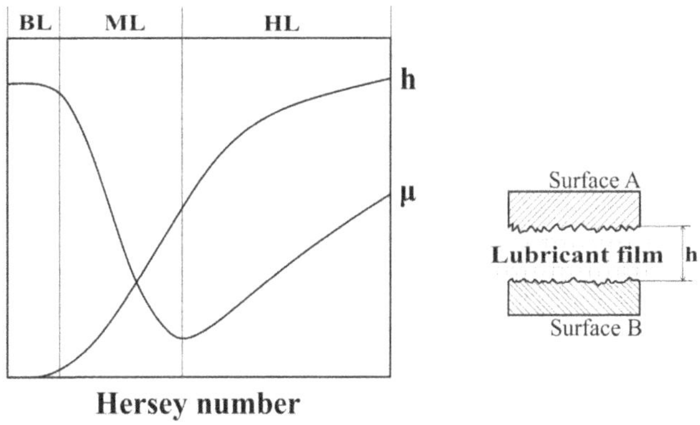

Figure 4.3. Modified Stribeck's curve.

4.2.4.1.1 Boundary-layer lubrication (BL)

In the BL regime, very close contact of surface asperities occurs with a very thin layer of lubricant. Therefore, the entire load is carried by the asperities itself. This regime occurs due to heavy load, lower sliding velocity, higher surface roughness or due to insufficient supply of liquid lubricant. The value of friction coefficient and energy losses due to friction are at peak on account of asperity contact, which further results in the heat generation at the interface [37].

4.2.4.1.2 Hydrodynamic lubrication (HL)

There is no contact between surface asperities, both surfaces are separated by thick film lubrication. Therefore, it is also referred to as full film lubrication and the most desirable lubrication regime under minimal loading conditions. However, HL fails at low lubricant viscosity, shock loading, and low-speed high torque conditions. The lubricant employed should withstand a wide range of temperature and pressure changes as fluid friction occurs between internal layers of lubricant [38, 39].

4.2.4.1.3 Mixed lubrication (ML)

It is sometimes termed as an intermediate lubrication regime between BL and HL. The value of the frictional coefficient decreases as a result of reduced asperity contact and increased film thickness compared to BL. The load is partially shared by few contacting asperities while the rest of the load is carried by lubrication film [40, 41].

While selecting a lubricant for the desired application, its performance characteristics, such as VI, resistance to oxidation, temperature changes, flash point, pour point, and lubrication regimes are taken into consideration. VI is the measure of fluid viscosity relative to the change in temperature of the working environment. It determines the frictional losses and thickness of fluid film regimes. Higher VI of a fluid signifies limited alteration in the fluidity concerning temperature changes. In an ICE, high VI lubricant oil shows reduced oil and fuel consumption with less wear [42–44]. The main three types of liquid lubricants: mineral oil, synthetic oil, and vegetable oils, are briefly described with their resources and generic properties.

4.2.4.1.4 Mineral oils

Mineral oils are derived from crude oil at a refinery and distinguished by their chemical structure as paraffinic, naphthenic, or aromatic. The majority of the commercial lubricant market is influenced by mineral oils as they are inexpensive and ubiquitous. Mineral oils are obtained from crude oil by the fractional distillation process. These hydrocarbons possess three different chemical structures, viz. paraffinic, naphthenic, and aromatic. Paraffin is long straight-chain hydrocarbons with a high pour point on account of wax content. The pour point is the lowest temperature below which liquid loses its flowability. On account of high viscosity index (VI) and low specific gravity, these are mostly utilized in engine oils, transmission fluids, and gear oils. However, naphthenic oils holding low VI, high specific gravity, and lower pour points are applied in rubber process oils, metal-working oils, industrial greases, and electrical transformer oils [35, 45, 46].

4.2.4.1.5 Synthetic oils

Depending upon the operating conditions, synthetic oils are artificially structured using uniform hydrocarbon molecules. Their molecular structures are designed and controlled relative to the performance characteristics for desired applications, such as oxidation stability at higher temperatures or low pour points to retain flowability, even at subzero temperatures. Polyalphaolefin (PAO), polyalkyleneglycol (PAG), and synthetic esters (SE) are some of the most common synthetic oils having widespread applications in the lubrication industry. Reduced intermolecular interactions in PAO prevent crystallization at a lower temperature to retain fluid flow. PAGs hold high VI, better control over friction and sedimentation (sludge). SE being non-toxic and biodegradable in nature bears the capacity to withstand high loads, high shear forces. Apart from SE, the majority of synthetic oils are toxic, non-biodegradable, and expensive [46, 47].

4.2.4.1.6 Natural oils

Lubricant discharge in the environment after its use provoked serious concern about the biodegradability of the lubricants. The extensive demand for base oils for lubrication cannot absolutely rely on nonrenewable sources. Considering limited mineral resources from Earth's deposits, the consumption rates are significantly higher to seek out renewable, eco-friendly alternatives. Bio-lubricants, derived from natural resources such as vegetable oils and animal fats are gaining popularity owing to their superior lubrication properties along with renewability, biodegradability, eco-friendliness, and non-toxic nature [48, 49]. Natural oils are comprised of triacylglycerol acquired from vegetables, fruits, seeds, or animal fats. Triacylglycerol is a polar fatty acid that promptly reacts with the metal surfaces to form a thin layer. The layer obstructs asperity contact, which ultimately results in reduced friction, enhanced boundary lubrication, and anti-wear properties [50]. Bio-based base oils are utilized to formulate bio-lubricants for applications where the loss of lubricant in the eco-system is inevitable, such as in mining, agriculture, food processing, forestry, outboard motors, and marine industries [51].

4.2.4.2 Bio-lubricants

Biodegradation is an aerobic disintegration of organic matter attributed to microbial activity. As mentioned earlier, base oils count for the major part of lubricant constituents, therefore, lubricant biodegradability relies heavily upon the basestock. Bio-lubricants can be comprised of bio-based oils, or sometimes SE-based products. Bio-based oils draw oils from different plants such as soybean, canola, avocado, peanut, tallow, lard, canola (rapeseed), castor, coconut, corn, cottonseed, olive, palm, safflower, sesame, and sunflower or from animal fats. Bio-lubricants derived from these oils are biodegradable and non-toxic having high flash point and sustainable basestock and revealing better tribological properties than mineral oil-based lubricants. Nevertheless, these bio-lubricants are susceptible to oxidation at higher temperatures and possess high pour points, which induce further structural modification. Genetic alteration in sunflower and canola (rapeseed) oil is carried out, which increases oleic acid content to achieve oxidation stability [52–55].

SE are glycerol molecules obtained from plant and animal sources. However, these esters possess unstable structures corresponding to bio-based oils, hence further esterification from a series of alcohols and acids provides a stable molecular structure. Moreover, the synthesis offers access to a diverse range of alcohols and acids to formulate a lubricant with a broad set of operational characteristics. The characteristics incorporate low pour point, high flash points, and enhancement in thermal and oxidation stability [46]. Withdrawal of a hydroxyl group from esters results in the alteration of molecular structure, which lads to improved VI of SE [56]. The bio- lubricants derived from tilapia (fish) oils revealed excellent thermal stability owning to the longest chain alcohol [57]. Introduction of triesters in Karanja oil structure and high oleic acid content improves tribological characteristics and flash point for automobile application [58].

4.2.4.3 Ionic liquids

A modern class of environment-friendly lubricants known as ionic liquids (ILs) is gaining popularity due to their beneficial lubricant characteristics and their derivation from bio-based feedstocks. These are liquid salts containing organic anion–cation moieties linked through a variety of bonds in a lamella and the whole structure looks like a solid crystal structure where each lamella is holding together, owing to Van der Waals forces. ILs offer constant lubricating film between the contacting surfaces on account of negatively charged anions that adhere to the positively charged metal surfaces. Corresponding to the adsorbed monolayer, ILs offer remarkable enhancement in tribological properties, exhibiting high thermal stability, low melting point, liquidity over wide temperature ranges, and minimal vapor pressure [59–61].

4.2.4.4 Lubricant additives

In high-temperature applications, lubricants are prone to drop their viscosity which leads to deleterious consequences during operating conditions. Therefore, thickeners are utilized in lubricants which serve a thickening effect by expanding their molecules at higher temperatures. These are termed as viscosity modifiers [62]. Sometimes lubricant fails to function where extraneous objects such as dust, water, or soot contaminate the lubricant which further results in abrasive wear at the contacting surfaces. Therefore, different types of anti-wear agents and extreme pressure (EP) agents are added to lubricants [63]. Moreover, water molecules foster corrosio of surfaces, while partially burnt fuel affects lubricant viscosity. To overcome such adverse consequences and enhance performance, lubricants are provided with additives depending upon their functionality [64, 65]. Detergents, dispersants, anti-oxidants, inhibitors, anti-foam agents, emulsifiers, demulsifiers, and pour point depressants are some of the different types of additives used in lubricants [66–70].

4.2.4.4.1 Nanoparticle additives

Presently, widespread nano-technology is attaining advancement in numerous fields, thereby improving the effectiveness of lubricants by compounding nanoparticles as additives in them. Nano-sized particles encourage lubrication by stockpiling between the asperities of the contacting surfaces, subsequently reducing the friction and wear phenomenon. Such a smooth surface is obtained by the polishing effect mechanism, while sometimes these nanoparticles get accumulated into the grooves of an abraded surface under the significance of the mending effect mechanism. Furthermore, when the nanoparticles act as a rolling medium or form an amorphous layer between the surfaces to minimize contacting area, it is termed a ball-bearing mechanism or protective film mechanism, respectively [71–74]. Some of the frequently used nanoparticle additives and their influence on tribological properties of contacting surfaces are shown in table 4.1.

4.3 Tribological behavior of mechanical components in ICE

Presently, in the transportation sector, ICEs are accountable for the source of power generation. Therefore, researchers are incessantly putting efforts into improving the performance of these engines. The majority of mechanical power loss in an ICE is ascribed to frictional losses, thus, significant attention has been drawn towards reducing friction in order to enhance the mechanical efficiency of an ICE. In an operating condition, several mechanical components of an ICE are set in rolling and sliding motion relative to corresponding surfaces. These surfaces in contact often experience friction on account of dynamic loading, which results in straining, heat generation, and material loss at the interface. The power generation in an ICE is executed through the expansion of combustion gases in the combustion chamber subjected to fuel burning. The fuel is drawn into the combustion chamber through valves which are operated by a cam and tappet mechanism. The expanded combustion gases displace the piston to produce a power stroke which further transforms into the rotation of the crankshaft. Subsequent power distribution is carried away by the gear mechanism under the loading conditions. The gear drive accomplishes the various combinations of speed and torque requirements. The demand for high-speed engines for long-term applications impels ICEs to endure innumerable cycles without failure. The mechanical components confront an immense amount of stress which leads to wear and fatigue failures. For sustenance of such components, their tribological advances need to be understood [92–95].

Table 4.1. Frequently used nano-scale lubricant additives [75].

Reference	Nano-scale additives	Base oil	Attributed findings	Application
[76]	Tungsten disulfide (WS_2)	Synthetic oil 5W-30	High density and large specific surface area induce mending effect reducing wear scar diameter (WSD) to show anti-wear and anti-friction characteristics	Engine lubrication
[77]	Titanium dioxide (TiO_2)	API-1509 Engine oil	Dispersion stability displayed a reduction in friction and wear	Gear lubrication and industrial lubrication, particularly in cutting operation
[78]	Molybdenum disulphide(MoS_2) nanosheet	Hydraulic oil	Chemical absorption forms solid lubricant, dropping coefficient of friction to 0.04	Sliding bearings of electro hydrostatic actuator (EHA) of an aircraft
[79]	Hexagonal-boron nitride (h-BN)	Synthetic oil SAE 20W50	Reduced WSD and wear loss up to 20% and 70%, respectively	Piston ring and cylinder liner lubrication
[80]	h-BN	Semi-synthetic(SE 15W-40)	Improved anti-friction and anti-wear	As a lubricant to dissipate heat in high friction environments
[81]	Silica (SiO_2)	An aqueous solution of polyethylene glycol-200	Improved anti-friction characteristic of a lubricant ascribed to mending effect	Water-based lubricants
[82]	SiO_2 particles	Semi-synthetic oil 5100 4T Hybrid 10W30	Reduction in surface roughness and WSD resulted in decreasing engine noise and scuffing	Motor lubrication
[83]	Graphene	Synthetic oil SAE 10W30	Reduced surface roughness consequently reduces friction coefficient and specific wear rate.	Engine lubrication

(*Continued*)

Table 4.1. (*Continued*)

Reference	Nano-scale additives	Base oil	Attributed findings	Application
[84]	Aluminium oxide (Al_2O_3)	Polyalkylenegycol (PAG)	Increased thermal conductivity along with viscosity threshold	Automotive air conditioning system
[85]	Multi-walled carbon nanotubes (MWCNT)/MgO	Motor oil SAE50	Enhanced dynamic viscosity threshold	Heavy-duty diesel engine oil
[86]	Copper oxide (CuO)	Mineral oil	Deposition of nanoparticles at worn out surfaces showed 50% reduction in frictional coefficient in mixed lubricant regime	Marine engine lubrication
[87]	Cerium oxide (CeO_2)	Ethylene glycol	Increased thermal conductivity	Coolants
[88]	Zinc oxide (ZnO)–silicon nitride (Si_3N_4)	Lithium lubricating grease	Friction coefficient and WSD reduced by 34.3% and 8.3%, respectively followed by a reduction in vibration and noise.	High-speed bearings
[89]	Al_2O_3/TiO_2 hybrid nanoparticles	Synthetic oil 5W-30	Decrease in COF and wear rates by 47.61% and 17%, respectively	Piston ring and cylinder liner lubrication
[90]	ZnO/graphene nano-composites	synthetic ester lubricant (SparkM40)	Excellent load distribution and stable lubricating film	
[91]	Copper/ polydopamine (PDA)/MoS_2 nano-sheets	Sunflower oil	The friction coefficient and wear scar width were decreased by 37.8% and 28.5% compared to neat sunflower oil	

4.3.1 Gears

4.3.1.1 Terminology

A gear is a circular toothed mechanical component that transmits the motion as well as the power to another gear on account of its rotation. Depending on the axis of rotation, gears provide transmission of motion in multiple directions, while several other design parameters, such as gear module, determine the power transmission at a variable rotational speed. Figure 4.4 illustrates a generic gear tooth where the gear module is the ratio of pitch circle diameter to the number of teeth that gear possesses. Gear module defines the size of the gear, while gear terminology helps to understand briefly the classification, working, and occurrence of failure criteria in a gear-pair [9, 96–99].

4.3.1.2 Classification

Gear-pairs are categorized on the basis of their axis of rotation and translation of motion in different planes. Spur gears, helical gears, internal gear-pair, and gear rack are parallel axes gear-pairs that rotate in the same plane. Bevel or spiral bevel gears operate in different intersecting planes whose axis of rotation also meets at some point. Screw or worm gears are the type of gear-pairs whose axes of rotation never meet and also operate in different planes [100].

4.3.1.3 Working

Gears are capable of generating variable torque at different rotational speeds and can transmit power in different directions. Consider a simple gear-pair consisting of two gear wheels, where one is called driving gear which provides input from some

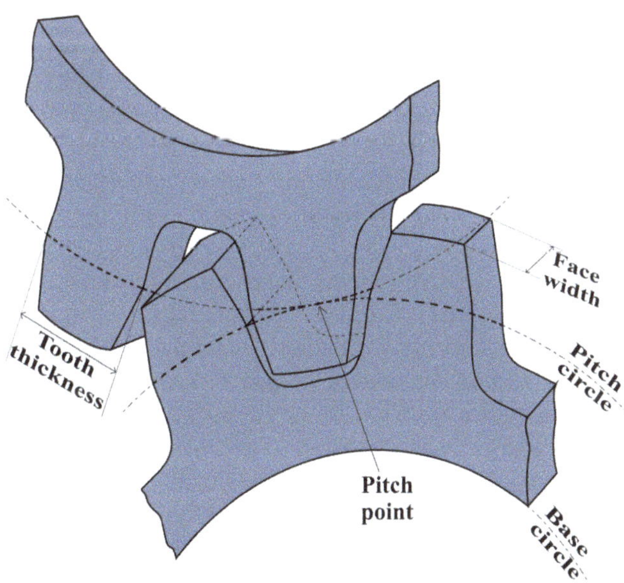

Figure 4.4. Gear terminology.

power generating devices such as an ICE shaft or a motor. Another wheel is termed a driven gear, whose function is to transmit the output power from the driving gear. One can either increase power output in the form of torque by providing a greater number of teeth to the driven gear compared to the driving gear or increase the rotation speed on output by decreasing the number of teeth on a driven gear. The driven gear always rotates in the direction opposite to that of the driving gear. If the driving gear is rotating in a clockwise direction, the output must be in a counter-clockwise direction. However, the axis of rotation can be varied by employing the different types of gear-pairs mentioned earlier [101–103].

4.3.1.4 Failure

Gears are considered to be the efficient power transmitting mechanism compared to belt drives or chain drives as they prevent slippage when meshed properly. However, gears might fail after functioning for their prolonged service life. The gear does not entirely break down all of a sudden, it usually retains its complete functionality for longer runs with a few damaged teeth. Regarding their working conditions, gears are subjected to repetitive loading, which leads to fatigue failure. Generally, an enmeshed tooth in a gear-pair makes a line contact between them. The gear teeth experience both rolling as well as sliding friction at their contacting surfaces during load transfer from one gear to another. During the course of action, frictional energy gets dissipated in the form of heat and plastically deforms the micro-asperities causing surface wear [104–106]. Pitting is the most common failure phenomenon observed in gears as a result of repetitive contact stresses surpassing the surface fatigue strength of the material. Materials at the contacting surfaces get displaced, forming a micro-pit in a fatigue region. These micro-pits further spread deformities over the tooth surfaces as stress concentration continues to increase during the cyclic loading. Further dispersion of the micro-pits induces cracks where usually lubricants seap in, causing scarcity of lubrication at the interface, which severely corrodes the gear tooth surface. Deleterious repercussions of this progressive pitting sometimes get hampered at early stages on the grounds of rolling action [107, 108]. Sometimes gear surfaces experience corrosion owing to the presence of moisture or chemicals. Pitting corrosion wears out the surface leaving a significant amount of wear debris to contaminate the lubricant. Wear debris also accounts for the quantitative loss of gear material [109, 110].

Scuffing occurs when excessive loading degenerates lubricant fluid film between contacting surfaces in the BL regime. Further increase in friction and heat generation results in momentary localized welding of asperities in contact. Subsequent rolling or sliding movement of gears shears apart the glued surfaces. This worn out layer leaves a scuff on material surfaces, making it prone to more severe failure phenomena such as corrosion, crack propagation, seizure, scoring etc [111–113]. Failure in gear-pairs is avoided by selecting appropriate gear material with anti-wear properties and optimized design configuration. Working conditions, type of machining process, post-treatments, and lubrication also play an important role in determining the endurance strength of the gear-pair. Iron-based alloys as a gear material are a popular choice due to their mechanical properties, which are further enhanced by employing treatments such as coating, shot peening, or ultrasonic surface rolling [114]. The magnitude of the friction coefficient is diminished by employing coating technology to the gear material. An increase in the stiffness of coating material resulted in the enhancement of fatigue strength along with the durability of a gear-pair [115]. Almost 80% decrease in wear rates along with a decrease in friction coefficient was observed when 17Cr2Ni2MoVNb steel was treated by the shot peening process, whereas the same material being treated with ultrasonic surface rolling resulted in an increment of micro-hardness along with endurance against fatigue failure [116, 117]. Gear oil lubricant compounded with anti-scuffing additives such as titanium dioxide TiO_2 or Ni–MoS_2 nano-composites led to a diminution in friction and wear phenomenon. [77, 118]. Yttrium oxide Y_2O_3-based zirconium dioxide (ZrO_2) additive showed wear resistance application in extremely high-pressure conditions [119]. The fiber laser technique has been applied to create micro-pits on the TiN-coated $42CrMo_4$ steel surface. Sedimented wear debris got trapped into such pits enhancing resistance to wear by 70% subject to volume and 45% subjected to wear depth under high loading conditions. Furthermore, a decline in friction coefficient was also observed in tungsten carbide/carbon (WC/C) coating [9, 120].

Bio-lubricant consisting of 15% fatty acids acquired from *Calophyllum inophyllum* seeds combined with a commercial gear oil displayed advancement in load-bearing capacity of a lubricant with decreased friction and wear rates [121]. Epoxide soyabean oil blended with synthetic esters also displayed improved wear properties of gear oil [122]. The addition of ILs to the synthetic ester-based bio-lubricant of wind turbine gears operating at high speeds resulted in a decrease in wear losses by 74.3% compared to pure base oil [123]. Biodegradable lubricants were formulated using TiO_2 and CuO nanoparticles in sunflower and palm oil for worm gear. 0.2% of TiO_2 and CuO in sunflower oil showed a decline in friction coefficient by 51% and 48%, respectively, compared to the conventional gear oil. The reduced friction was attributed to the deposition of nanoparticles forming a tribofilm on the effective surface of worm gear [124].

4.3.2 Cams and tappets

Usually, most ICEs use a cam-operated mechanism to govern the motion of inlet and outlet valves. Cams and tappets are both parts of the valve-train assembly which also regulates the operational timings of valves in ICEs. However, in some modern ICEs, the cam and tappet mechanism drives fuel injector and oil pumps as well.

4.3.2.1 Terminology

A cam has irregular contours according to the motion it is supposed to transmit through the follower. In ICEs, a circular cam exhibits eccentricity at some point, as depicted in figure 4.5. Cam lobes are mounted camshafts at certain positions corresponding to the position of cylinders. Cams are employed to transmit the rotational motion of camshafts into the translatory motion of tappets to operate valves inside the combustion chamber. A tappet provides a consistent footing for the

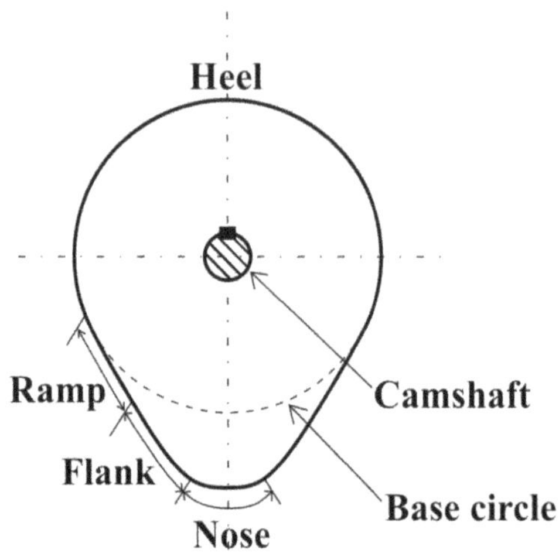

Figure 4.5. Cam terminology.

valve to operate in a linear motion derived by following a cam lobe profile. Tappets are held against the cam surfaces on account of spring force or hydraulic pressure in the case of the hydraulic tappet. Spring force exerted on a cam surface is constant throughout the heel area, which tends to increase as the tappet follows the cam ramp. The valve appears to open as the tappet further trails on the cam flank surface. Spring force is at its maximum on a cam nose surface as the tappet spring is entirely compressed against it [125–128].

4.3.2.2 Classification

There are varieties of tappet designs available, yet these are broadly classified into two categories depending on the type of frictional motion (sliding or rolling) tappets are confronting relative to cam surface. Sliding tappets are usually provided with some radius to minimize the contacting surface area. In accordance with the tappet configuration, cam profiles are also adapted for sliding tappets where cam lobes are made more tapered. However, for the rolling tappets, cam profiles are more rounded which signifies the opening of the intake valve for an extended time owing to excessive contact forces; rolling tappets experience roller slippage and significant frictional losses [129–131].

Early ICE designs implemented an overhead valve (OHV) for its functioning. The valve-train mechanism encompasses cam, tappet, pushrod, rocker arm, and valve. A tappet situated down in the engine block transmits linear motion derived from the rotating camshaft to the pushrod. The pushrod further triggers the rocker arm to convey movement to the valves located at the cylinder head. Considering all the moving components, this supplementary inertia led to increased vibrations and, subsequently, increased chances of failure [132–135]. Later on, ICEs adapted overhead camshaft (OHC) engines to reduce undesirable outcomes generated by OHV engines. Here, the camshaft is situated in the cylinder head above the combustion chamber, eliminating pushrod requirement. The valve is directly connected to the tappet while maintaining its contact with the cam on account of spring action [136].

OHC engines are further classified into a single overhead camshaft (SOHC) and double overhead camshaft (DOHC). DOHC engines utilize two camshafts, which allocate provision for multiple valves for inlet and outlet as well. Multi-valve OHC engines deliver high horsepower for high-performance vehicles. Precision in valve timings at higher rpm makes these engines more efficient compared to OHV engines [137, 138].

4.3.2.3 Working

In an ICE, crankshaft output drives the camshaft rotation to operate valves on cylinder heads through the valve-train assembly. A tappet follows the eccentric cam profile on account of relative rolling and sliding, transforming the rotational movement of the cam into the translatory motion of the tappet. The movement of the tappet is perpendicular to the rotational axis of a camshaft. As the cam rotates, the tappet starts moving away from the cam center, simultaneously pushing the valve stem. The opening of the valve is initiated through the tappet rise when the cam pushes the tappet against the spring. Eventually reaching towards the most eccentric point on a cam, the valve opens completely along with the spring retraction. Ongoing rotation of cam brings tappet return to cam's minimum radius profile parallelly closing the valve by means of spring force [139–142]. In practice, cam and tappet operate at very high speeds, the camshaft speed of rotation in four-stroke ICEs is half of the crankshaft's speed. Therefore, any negligence in its design, lubrication, or minute defacement is most likely to develop noise and vibrations, which further leads to failure while functioning [143]. Sometimes hydraulic tappets are deployed for their longevity, quiet and smoother operations. It makes use of fluid pressure to replace spring force [144].

4.3.2.4 Failure

Spring force, inertia force, and frictional force developed in cam and tappet are sometimes exclusively or inclusively accountable for their failure. During each return cycle from the crest to the minimum radius point on the cam profile, the tappet exerts a spring force on a cam surface. Excessive spring force causes localized stresses on a cam surface, while inadequate spring force also makes the tappet jump off the cam surface. At high speeds, rotational and linear movement of cam and tappet lead to inertia forces due to their mass. Excessive eccentricity leads to the increased linear displacement of the tappet, meanwhile, high cam rpm causes the tappet to impart more spring force along with the inertia force to curtail the fatigue life of the cam. The tappet retains the contact with a cam surface over the same line of points for a prolonged time on account of rolling and sliding friction [133, 145]. The contacting surfaces operate at high speeds inducing high temperature and drastic pressure changes, which further affect the lubricant viscosity. The occurrence of localized contact stresses and cyclic loading commenced crack propagation subjected to pitting. Cracking of lubrication film further prompts seizing and scuffing failure. These are rather similar failure criteria as perceived in gears owing to abrasive as well as adhesive wear. Creating a random structured pattern over tappet shim by shot peening process did not contribute to friction abatement but rather increased surface roughness due to the formation of micro-pits, whereas parallel V-shaped channels made by a diamond tool lower the friction [146]. Silicon nitride-based ceramic (Si_3N_4) is a broadly picked cam as well as tappet material. Adding rare-earth (RE) oxides to Si_3N_4 resulted in the improvement of surface roughness and wear resistance of the material [147]. Steel (Cr12W) roller tappet undergoes dry friction subjected to oxidation and adhesive wear when tested on steel (GCr15) surface. Then the roller material was treated with ion nitriding and wear mechanism transformed to spalling in lubricated conditions [148]. For the lubrication, at the cam and tappet interface, PAG oils initially displayed greater wear rates while almost 48% decrease in friction was observed when compared to GF-5 base oils. Furthermore, adding zinc dialkyldithiophosphate (ZDDP) in PAG showed a potential decline in wear [149]. The addition of 10% of waste cooking oil (WCO) biodiesel into mineral oil lubricant (SAE20W40) revealed a decrease in wear debris, friction, and wear rates [150].

Bio-based lubricants, such as palm oil containing trimethylolpropane (TMP) esters, have proved to be a better choice over polyalphaolefin (PAO) base oil for cam and tappet lubrication. Load-bearing capacities and operating temperature ranges of TMP are greater owning to the VI, which is 1.5 times greater than that of PAO. Friction torque at the contacting surfaces and the measure of wear at the cam nose surface are greater in PAO when compared to TMP. Moreover, addition of glycerol monooleate (GMO), molybdenum dithiocarbamate (MoDTC), and ZDDP as lubricant additives showed excellent friction and wear-resisting characteristics [151]. Mixing 10% of waste cooking oil (WCO) in a lubricant (SAE20W40) used for the cam–tappet valve-train system displayed minimized wear debris, which indicates reduced friction and wear rates [152]. Although the application of bio-lubricant for the cam–tappet system is limited, more research needs to be carried out.

4.3.3 Piston rings

Operating at full-load conditions, the temperature at the combustion chamber in an ICE reaches its maximum, causing piston material to experience thermal expansion. Therefore, pistons are deployed inside the cylinder bore, providing a requisite clearance between both surfaces. However, a significant clearance gap discharges fuel or burnt gases from the combustion chamber to contaminate the lubricant in the crankcase. Also, immoderate piston expansion causes engine seizure. To compensate for that clearance gap and prevent intermixing of fluids from the combustion chamber into the crankcase, sealing in the form of piston rings is deployed. Besides being an effective sealant, piston rings function as a medium to dissipate heat from the piston to the cylinder walls. It also regulates the lubricant flow between the piston and cylinder liners to minimize the friction between two contacting surfaces in sliding motion [153–155].

4.3.3.1 Terminology

Piston rings are split at one point on a circumference providing an end gap for thermal expansion during high working temperatures. This end gap also allows easy installation into the piston grooves. These rings get compressed when placed inside the cylinder bore as piston rings possess a slightly larger diameter than the bore diameter. The pre-tensioning in the rings tends to apply pressure against the cylinder liner surface, sealing the combustion chamber. Cylinder liners are a very thin-walled cylindrical components fitted inside a cylindrical block of the ICE. Cylinder liners served as a sliding surface for piston rings while retaining lubrication in between both surfaces [156].

Rectangular cross-section compression rings are the most commonly utilized ring, however, different cross-sections serve a variety of applications accompanied by the type of engine. Altering the cross-section of the compression piston ring eliminates the need for oil control rings at times. A few slots are pierced through the inner surface towards the periphery of an oil control ring to allow the flow of excessive lubricant scraped from the liner surface. The lubricant is then collected in the inner grooves of the ring and disposed back to the crankcase. Sometimes these rings are installed with coil springs for additional pressure to scrape out the lubricant in extreme lubrication conditions [157, 158].

4.3.3.2 Classification

Depending on the engine type, there are multiple rings installed into the grooves of the piston. Piston rings are classified into two categories, compression rings, and oil control rings, based on the purpose they serve. The compression rings wider than the cylinder bore diameter create pre-tension in the rings, which conform to the cylinder liner surface against the spring pressure providing absolute sealing [159].

Apart from sealing off the combustion chamber from the crankcase, there is a need to regulate lubricant oil between the contacting surfaces in motion to avoid frictional losses. Installed in the bottom-most groove of the piston, as displayed in figure 4.6, these rings evenly distribute the lubricant over the circumference of the cylinder liner. The function of the oil control ring is to provide and control the amount of lubrication in between contacting surfaces of the piston and cylinder liner. Oil control rings are absent in some engines as one of the bottom compression rings performs the function of the oil control ring [160–162]. Scraper or wiper rings are installed below compression rings whose function is to scrape off the excessive oil from the liner surface behind the piston down to the crankcase. Sometimes specially designed piston rings are deployed after the compression ring, known as Napier rings. They assist the function of compression rings by means of a narrow line of contact from the extended nose part. The combustion gases that escaped from the prior compression ring are trapped and restricted from flowing further down towards the crankcase. The cross-section of Napier rings in figure 4.7 shows the undercut called an accumulator groove which provides more volume to scrape off excessive lubricants during the downstroke [163, 164].

One of the supplementary functions of the piston rings is preventing the piston from slamming against the cylinder walls. At high speeds, a slight transverse

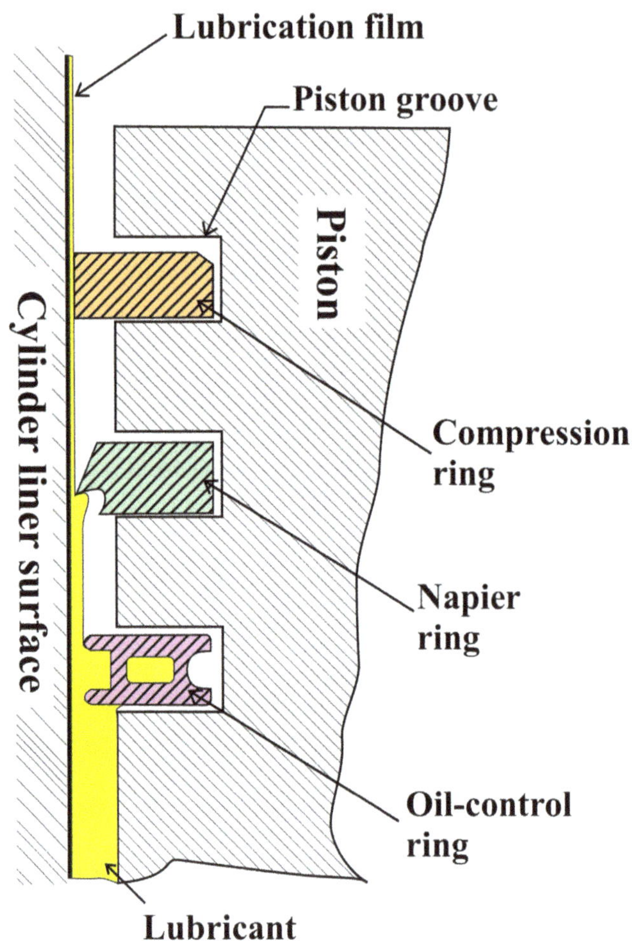

Figure 4.6. Schematic of piston ring operation.

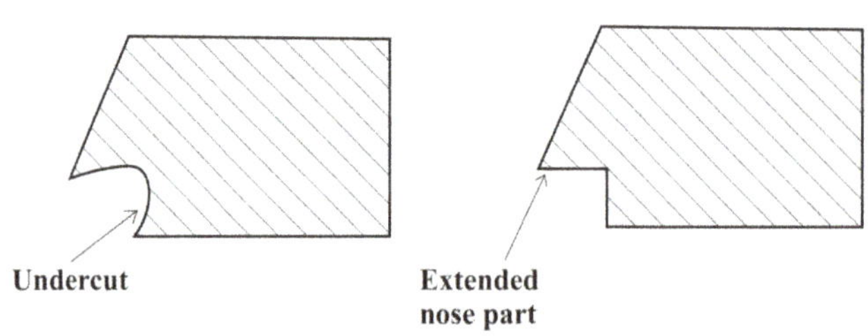

Figure 4.7. Napier ring cross-section.

movement of the piston due to excessive clearance between the piston and liner surface leads to detrimental consequences such as severe vibrations or instant failure. Piston rings hold the piston upright to retain its stipulated linear reciprocating movements.

4.3.3.3 Failure

Any deformity in the piston rings directly affects the ICE's efficiency as piston rings are located inside the combustion chamber where actual fuel burning takes place to generate power. Piston rings are nothing but dynamic seals sliding between piston and cylinder liner surfaces. This sealing encounters very high speeds, high temperatures, and drastic pressure changes. It also encounters inertia loads evoked from the acceleration and deceleration of piston movement inside a cylinder, damping offered by a lubricant film and frictional forces on account of sliding motion [165]. In almost all surfaces in relative motion, surface asperities tend to cause friction, adhesive and abrasive wear phenomenon. Surprisingly in piston ring and cylinder liner, asperities wear off during prolonged working and smaller reservoirs build up, which changes the lubrication regime from BL to HL [166, 167]. Spring pressure, along with the pressure generated from the expansion of burnt gases pushes the rings into the piston grooves and against the liner surface firmly. Piston rings are allowed to move freely in the piston grooves to retain close contact with the liner surface during periodic movements of the piston. High-temperature combustion gases from the chamber tend to raise the temperature of the piston as well as the piston rings, affecting the lubrication film between both surfaces. On account of the blow-by phenomenon, momentary starved lubrication produces a window for high-pressure gases to surge past the piston rings and contaminates the lubricant in the crankcase [168, 169]. Scratching and slapping of the piston on a cylinder wall caused by any slight disorientation in the linear reciprocating piston movement leads to abrasive wear and undesirable vibrations in an ICE. Piston rings hold up the piston tight against the cylinder walls to avoid such a phenomenon [170].

Lubricant film thickness and surface roughness determine the tribological behavior of piston rings and cylinder liner surfaces [171]. Smoothening down the cylinder liner surface provided with low viscosity lubricants has reduced losses in the engine due to friction and wear, subsequently minimizing the CO_2 emission [172]. Certain textures were structured on cylinder liner surfaces by honing the process to reduce friction between liner and piston ring interface. Reduced honing angle below 55° resulted in a decrease in the frictional coefficient of grey cast iron cylinder liner surface [173]. Diamond-like carbon (DLC) coated piston rings with smoothened plateau liner surface served better tribological properties than a honed surface for a diesel engine in extreme loading conditions [174, 175].

A significant depletion in friction was observed when laser surface texturing was employed to enhance lubrication by producing micro-pockets on the outer surface of the piston rings [176]. Electroplated chromium-diamond coating and chromium nitride (CrN)—graphite-like carbon (GLC) coating, among six other coating techniques for piston rings, have shown enhanced anti-scuffing and the ability to withstand high pressure and loading conditions [177]. A set of piston rings were coated with a combination of diamond-like carbon and tungsten carbide (DLC and WC). Later, they were coated with a graphene nanoparticle-reinforced epoxy polymer composite compounded in lubricant base oil SN150. The double coating layer prevents the piston rings from wear in extreme stress and temperature conditions [178]. Piston ring coating comprised of MoS_2–graphene oxide (GO) particles diffused in nickel–phosphorus (Ni–P) matrix has shown superior wear resistance and anti-friction characteristics among three other Ni-based coatings [179]. Hybridized nanoparticles alumina (Al_2O_3) and titanium oxide (TiO_2) in synthetic oil (5W-30) form a protective layer of solid lubricant, reducing friction, wear, and scuffing between piston rings and cylinder liner interface. Moreover, the addition of oleic acids to pure lubricant oil decreases the COF by 18% [89].

When tested in the similar working environment as that of diesel contaminated commercial synthetic lubricant (SAE 20 W40) for cylinder liner and piston ring lubrication, bio-lubricant drawn from rapeseed oil displayed a tremendous decrease in wear up to 91.5%. Such a drastic decline in wear was a consequence of long-chain fatty acids and the polarity of palm oil methyl esters, which develop a coating over metal surfaces to protect them from wear. As a result of this coat, there was a significant drop in the coefficient of friction and wear debris which subsequently improved the lubricity of bio-lubricant [180]. In another study, maintaining similar operating conditions as a synthetic lubricant, the friction coefficient of bio-lubricant derived from rapeseed oil was 23% less due to the protective coat developed by the combined effect of ester polarity and long-chain fatty acids existing in the bio-lubricant [181]. A similar phenomenon was observed when bio-lubricant derived from cottonseed oil showed increased lubricity compared to that of synthetic lubricant SAE40. A reduced coefficient of friction was observed between the surfaces of piston rings and cylinder liner by virtue of a protective layer developed by fatty acids that existed in TMP esters of cottonseed oil [182]. In addition to the organic film formation over the metal surfaces of cylinder liner and piston rings, different combinations of Ni–P, MoS_2 and GO coatings were applied to minimize the friction and deleterious effects of corrosive wear. Bio-lubricant procured from rice husk has synergically reacted with the coating elements to form an excellent tribological interface to prevent metal surfaces from wear caused by corrosion, adhesion and delamination [183]. 0.5% of CuO nano-scaled particles were added to chemically modified high-grade rapeseed oil, which exhibited excellent wear resistance along with improved thermo-oxidative stability and lubricity of a lubricant [184].

4.4 Challenges and future scope of bio-lubricants for ICE components

Bio-lubricants lack oxidation stability, which indicates that lubricant is more susceptible to oxidizing rapidly in working conditions. This adversely affects the lubricity of a bio-lubricant which is undesirable for the proper functioning of ICE components. A higher fraction of bio-lubricants resulted in corrosive and adhesive wear since oxidized lubricant working at higher temperatures leaves corrosive acids behind, generating rough surfaces. Whilst chemical alteration in bio-lubricants without impairing their physiochemical properties and utilization of several different types of additives would resolve the problem, it simultaneously increases the cost [152, 182]. Although the introduction of nanoparticles has improved the performance of bio-lubricant, a higher fraction of these nanoparticles than the optimum percentage could result in increased wear as a consequence of increased interaction between solid particles, which further reduces lubricity [184]. The high pour point of bio-oils failed to operate the majority of bio-lubricants at subzero temperatures. Another limitation of using bio-lubricant is seasonality which affects the quality of feedstock and, subsequently, the effectiveness of a lubricant. Furthermore, the feedstock of bio-lubricant has to be acquired from non-edible oils as utilization of edible oils could increase the demands and price and further disrupts the social balance [61]. For heavy loading conditions such as in gear lubrication, EP additives are utilized to maintain fluid film thickness between the contacting surfaces. These EP additives hinder the biodegradability of a lubricant [185]. When chemically altered rapeseed oil-based bio-lubricant was applied, abrasive wear was observed on cylinder liner surfaces which were 12% higher compared to the synthetic lubricant (SAE20W40) [181].

Global rapid industrialization and development in the automobile sector have prompted an ever-growing demand for lubricants for machinery and ICEs. Considering constantly increasing crude oil prices, limited petroleum resources, and environmental concerns of formulating them, synthesizing bio-lubricants for a variety of applications is gaining the attention of the majority of researchers. Therefore, reliable, economical, and sustainable feedstock for bio-lubricant formulation must be sought. To promote the utilization of bio-lubricants and eliminate dependency on petroleum resources, wastelands can be used for oil-crop cultivation. Initiatives to cultivate more non-edible oil-growing plants are necessary to be taken as edible oil, and mineral oil sources are not sustainable. To attain more thermo-oxidative stability in bio-lubricants more chemical alteration techniques are required to be sought. Although synergic effects of bio-lubricants with various types of coatings have shown incredible results, more variants need to be explored as current coating techniques such as DLC coating increases the overall cost of the lubricant. The performance of bio-lubricant with several different combinations of nanoparticles needs to be explored, considering its high cost. Few mechanochemical treatments, such as microwaves, ultrasound, and oscillatory flow reactors, have been established to improve production yield and curtail energy usage.

4.5 Conclusions

The quantum of tribology incorporating the occurrence of friction and wear together with the influence of lubrication have been studied. Friction degrades the surfaces in relative motion, while the wear mechanism accounted for material loss. Various lubrication conditions, their classification, and different types of lubricant additives are reviewed. Lubricant performance can be optimized by selecting a combination of base oils and additives corresponding to the operational conditions of the desired application. The study also contributed to failure criteria subjected to the working terminology of engine components. The tribological behavior of gears, cams, tappets, and piston rings suggested that losses due to friction and wear were minimized by applying a coating or providing essential lubrication.

Considering the global environmental concern, the utilization of bio-lubricants in engine components appears to be an effective substitute owing to their renewable feedstock, biodegradability, and eco-friendliness. Several bio-based lubricants derived from natural resources applied to engine components and their outcomes have been discussed. The performance of bio-lubricants is enhanced by chemical alterations and the introduction of various additives. Implementation of nano-particle additives and surface coating techniques have shown significant enhancements in lubrication performance. Several types of research are still ongoing to improve the thermo-oxidative stability of bio-lubricants to entirely surpass synthetic lubricants.

References

[1] Hutchings I and Shipway P 2017 *Tribology: Friction and Wear of Engineering Materials.* (Oxford: Butterworth-Heinemann)

[2] Bhushan B 2013 *Introduction to Tribology.* (New York: Wiley)

[3] Majumdar B C 2008 *Introduction to Tribology of Bearings* (New Delhi: S. Chand Publishing)

[4] Czichos H 1986 Introduction to friction and wear *Friction and Wear of Polymer CompositesComposite Materials Series* (Amsterdam: Elsevier) pp 1–23

[5] Affatato S and Grillini L 2013 Topography in bio-tribocorrosion *Bio-Tribocorrosion in Biomaterials and Medical Implants* (Cambridge: Woodhead Publishing) pp 1–22a

[6] Dowson D 1998 *History of Tribology* 2nd edn (New York: Wiley)

[7] Affatato S, Spinelli M, Zavalloni M, Mazzega-Fabbro C and Viceconti M 2008 Tribology and total hip joint replacement: current concepts in mechanical simulation *Med. Eng. Phys.* **30** 1305–17

[8] Mang T and Dresel W (ed) 2007 *Lubricants and Lubrication* (New York: Wiley)

[9] Wang Q J and Chung Y W 2013 *Encyclopedia of Tribology* (Berlin: Springer)

[10] Khaemba D N, Jarnias F, Thiebaut B, Neville A and Morina A 2017 The role of surface roughness and slide-roll ratio on the decomposition of MoDTC in tribological contacts *J. Phys.* D **50** 085302

[11] Ranganathan N, Lawrence F A, Rajkumar S, Bensingh R J, Kader M A and Nayak S K 2020 Influence of surface roughness on tribological and mechanical properties of micro-milled and laser ablated poly (methyl methacrylate) PMMA organic glass *Polym. Test.* **81** 106184

[12] Arnell D 2010 Mechanisms and laws of friction and wear *Tribology and Dynamics of Engine and Powertrain* (Amsterdam: Elsevier) pp 41–72

[13] Lushnikov N and Lushnikov P 2017 Methods of assessment of accuracy of road surface roughness measurement with profilometer *Transp. Res. Proc.* **20** 425–9

[14] Szeri A Z 2010 *Fluid Film Lubrication* (Cambridge: Cambridge University Press)

[15] Ştefănescu D M 2020 Application of electromagnetic and optical methods in small force sensing *Handbook of Force Transducers* (Cham: Springer) pp 41–50

[16] He D, Pu J, Lu Z, Wang L, Zhang G and Xue Q 2017 Simultaneously achieving superior mechanical and tribological properties in WC/aCnanomultilayers via structural design and interfacial optimization *J. Alloys Compd.* **698** 420–32

[17] Xu J, Yi D Q, Cui Q and Wang B 2019 Enhanced wear resistance of Ni/h-BN composites with graphene addition produced by spark plasma sintering *Acta Metall. Sin. (Engl. Lett.)* **32** 876–86

[18] Zang L B, Chen Y, Wu Y M, Ran L X, Zheng Y, You D L and Bi W Y 2020 Tribological behavior of AISI52100 steel after PC/MoS_2 lubricant surface modification *Strength Mater.* **52** 118–29
[19] Ji G, He X F and Liao H 2019 Tribological properties of surface modified Cu nanoflakes prepared by plasma assisted ball milling *J. Mater. Eng.* **47** 114–20
[20] Ludema K C and Ajayi L 2018 *Friction, Wear, Lubrication: A Textbook in Tribology* (Boca Raton, FL: CRC Press)
[21] Kragelsky I V and Alisin V V (ed) 2016 *Friction Wear Lubrication: Tribology Handbook* (Amsterdam: Elsevier)
[22] Pennestrì E, Rossi V, Salvini P and Valentini P P 2016 Review and comparison of dry friction force models *Nonlinear Dyn.* **83** 1785–801
[23] Li X, Dong M, Jiang D, Li S and Shang Y 2020 The effect of surface roughness on normal restitution coefficient, adhesion force and friction coefficient of the particle-wall collision *Powder Technol.* **362** 17–25
[24] Bahaaddini M, Hagan P C, Mitra R and Khosravi M H 2016 Experimental and numerical study of asperity degradation in the direct shear test *Eng. Geol.* **204** 41–52
[25] Lenard J G 2007 *Primer on Flat Rolling* (Amsterdam: Elsevier)
[26] Patil S K 2012 Regenerative braking system in automobiles *Int. J. Res. Mech. Eng. Technol.* **2** 45–6
[27] Stoehr T, Eisenberg B, Suchert E, Spikes H and Fan J 2011 Advances in Tribological Design of Poly (Alkyl Methacrylate) Viscosity Index Improvers (No. 2011-1-2123) *SAE Technical Paper*
[28] Molinari J F, Aghababaei R, Brink T, Frérot L and Milanese E 2018 Adhesive wear mechanisms uncovered by atomistic simulations *Friction* **6** 245–59
[29] Abdulqadir L B, MohdNor N F, Lewis R and Slatter T 2018 Contemporary challenges of soot build-up in IC engine and their tribological implications *Tribol.-Mater., Surf. Interfaces* **12** 115–29
[30] Wang D, Song D, Wang X, Zhang D, Zhang C and Araújo J A 2019 Tribo-fatigue behaviors of steel wires under coupled tension-torsion in different environmental media *Wear* **420** 38–53
[31] Hua N, Hong X, Liao Z, Zhang L, Ye X, Wang Q and Liaw P K 2020 Corrosive wear behaviors and mechanisms of a biocompatible Fe-based bulk metallic glass *J. Non-Cryst. Solids* **542** 120088
[32] Reeves C J, Menezes P L, Lovell M R and Jen T C 2013 Tribology of solid lubricants *Tribology for Scientists and Engineers* (New York: Springer) pp 447–94
[33] Johnson D (ed) 2018 *LubricationTribology, Lubricants and Additives.* (Norderstedt: BoD–Books on Demand)
[34] Menezes P L, Reeves C J and Lovell M R 2013 Fundamentals of lubrication *Tribology for Scientists and Engineers* (New York: Springer) pp 295–340
[35] Bart J C J, Gucciardi E and Cavallaro S 2013 Principles of lubrication *Biolubricants: Science and Technology* (Cambridge: Woodhead Publishing) pp 10–23
[36] Maru M M, Trommer R M, Cavalcanti K F, Figueiredo E S, Silva R F and Achete C A 2014 The Stribeck curve as a suitable characterization method of the lubricity of biodiesel and diesel blends *Energy* **69** 673–81
[37] Lhermerout R, Diederichs C and Perkin S 2018 Are ionic liquids good boundary lubricants? A molecular perspective *Lubricants* **6** 9

[38] Gropper D, Wang L and Harvey T J 2016 Hydrodynamic lubrication of textured surfaces: a review of modeling techniques and key findings *Tribol. Int.* **94** 509–29
[39] Uddin M S, Ibatan T and Shankar S 2017 Influence of surface texture shape, geometry and orientation on hydrodynamic lubrication performance of plane-to-plane slider surfaces *Lubr. Sci.* **29** 153–81
[40] Profito F J, Tomanik E and Zachariadis D C 2016 Effect of cylinder liner wear on the mixed lubrication regime of TLOCRs *Tribol. Int.* **93** 723–32
[41] Usman A and Park C W 2017 Numerical investigation of tribological performance in mixed lubrication of textured piston ring-liner conjunction with a non-circular cylinder bore *Tribol. Int.* **105** 148–57
[42] Esfe M H, Arani A A A and Esfandeh S 2018 Improving engine oil lubrication in light-duty vehicles by using of dispersing MWCNT and ZnO nanoparticles in 5W50 as viscosity index improvers (VII) *Appl. Therm. Eng.* **143** 493–506
[43] Ramasamy U S, Lichter S and Martini A 2016 Effect of molecular-scale features on the polymer coil size of model viscosity index improvers *Tribol. Lett.* **62** 23
[44] Muller H K 2019 *Fluid Sealing Technology: Principles and Applications* (Abingdon: Routledge)
[45] Bart J C J, Gucciardi E and Cavallaro S 2013 Lubricants: properties and characteristics *Biolubricants: Science and Technology* (Cambridge: Woodhead Publishing) pp 24–73
[46] Rudnick L R (ed) 2020 *Synthetics, Mineral Oils, and Bio-Based Lubricants: Chemistry and Technology* (Boca Raton, FL: CRC Press)
[47] Luna F M T, Rocha B S, RolaJr E M, Albuquerque M C, Azevedo D C and CavalcanteJr C L 2011 Assessment of biodegradability and oxidation stability of mineral, vegetable and synthetic oil samples *Ind. Crops Prod.* **33** 579–83
[48] Heikal E K, Elmelawy M S, Khalil S A and Elbasuny N M 2017 Manufacturing of environment friendly biolubricants from vegetable oils *Egypt. J. Petrol.* **26** 53–9
[49] Srivyas P D and Charoo M S 2018 A review on tribological characterization of lubricants with nano additives for automotive applications *Tribol. Ind.* **40** 594–623
[50] Reeves C J and Menezes P L 2016 Advancements in eco-friendly lubricants for tribological applications: past, present, and future *Ecotribology* (Cham: Springer) pp 41–61
[51] Owuna F J 2020 Stability of vegetable based oils used in the formulation of ecofriendly lubricants–a review *Egypt. J. Petrol.* **29** 251–6
[52] Cecilia J A, Ballesteros Plata D, AlvesSaboya R M, Tavares de Luna F M, Cavalcante C L and Rodríguez-Castellón E 2020 An overview of the biolubricant production process: challenges and future perspectives *Processes* **8** 257
[53] Luna F M T, Cavalcante J B, Silva F O N and CavalcanteJr C L 2015 Studies on biodegradability of bio-based lubricants *Tribol. Int.* **92** 301–6
[54] Reeves C J, Menezes P L, Jen T C and Lovell M R 2015 The influence of fatty acids on tribological and thermal properties of natural oils as sustainable biolubricants *Tribol. Int.* **90** 123–34
[55] Reeves C J, Siddaiah A and Menezes P L 2017 A review on the science and technology of natural and synthetic biolubricants *J. Bio-Tribo-Corros.* **3** 11
[56] Zulkifli N W M, Azman S S N, Kalam M A, Masjuki H H, Yunus R and Gulzar M 2016 Lubricity of bio-based lubricant derived from different chemically modified fatty acid methyl ester *Tribol. Int.* **93** 555–62

[57] do Valle C P, Rodrigues J S, Fechine L M U D, Cunha A P, Malveira J Q, Luna F M T and Ricardo N M P S 2018 Chemical modification of Tilapia oil for biolubricant applications *J. Clean. Prod.* **191** 158–66
[58] Sharma U C and Sachan S 2019 Friction and wear behavior of karanja oil derived biolubricant base oil *SN Appl. Sci.* **1** 668
[59] Reeves C J, Siddaiah A and Menezes P L 2019 Friction and wear behavior of environmentally friendly ionic liquids for sustainability of biolubricants *J. Tribol.* **141** 051604
[60] Reeves C J, Siddaiah A and Menezes P L 2017 Ionic liquids: a plausible future of biolubricants *J. Bio-Tribo-Corros.* **3** 18
[61] Singh S K and Savoy A W 2020 Ionic liquids synthesis and applications: an overview *J. Mol. Liq.* **297** 112038
[62] Martini A, Ramasamy U S and Len M 2018 Review of viscosity modifier lubricant additives *Tribol. Lett.* **66** 58
[63] Chen Y, Renner P and Liang H 2019 Dispersion of nanoparticles in lubricating oil: a critical review *Lubricants* **7** 7
[64] Menezes P L, Lovell M R, Kabir M A, Higgs C F and Rohatgi P K 2012 Green lubricants: role of additive size *Green Tribology* (Berlin: Springer) pp 265–86
[65] Rudnick L R (ed) 2017 *Lubricant Additives: Chemistry and Applications* (Boca Raton, FL: CRC Press)
[66] Wong V W and Tung S C 2016 Overview of automotive engine friction and reduction trends—effects of surface, material, and lubricant-additive technologies *Friction* **4** 1–28
[67] Kontou A, Taylor R I and Spikes H A 2021 Effects of dispersant and ZDDP additives on fretting wear *Tribol. Lett.* **69** 6
[68] Soleimani M, Dehabadi L, Wilson L D and Tabil L G 2018 Antioxidants classification and applications in lubricants *Lubrication Tribology, Lubricants and Additives* (Norderstedt: BoD - Books on Demand) pp 23
[69] Minami I 2017 Molecular science of lubricant additives *Appl. Sci.* **7** 445
[70] Braun J 2017 Additives *Lubricants and Lubrication* ed W Dresel (Hoboken, NJ: Wiley) pp 117–52
[71] Shafi W K, Raina A and UlHaq M I 2018 Friction and wear characteristics of vegetable oils using nanoparticles for sustainable lubrication *Tribol.-Mater., Surf. Interfaces* **12** 27–43
[72] Dai W, Kheireddin B, Gao H and Liang H 2016 Roles of nanoparticles in oil lubrication *Tribol. Int.* **102** 88–98
[73] Gulzar M, Masjuki H H, Kalam M A, Varman M, Zulkifli N W M, Mufti R A and Zahid R 2016 Tribological performance of nanoparticles as lubricating oil additives *J. Nanopart. Res.* **18** 223
[74] Martini A, Ramasamy U S and Len M 2018 Review of viscosity modifier lubricant additives *Tribol. Lett.* **66** 58
[75] Kotia A, Rajkhowa P, Rao G S and Ghosh S K 2018 Thermophysical and tribological properties of nanolubricants: a review *Heat Mass Transfer* **54** 3493–508
[76] Wu N, Hu N, Zhou G and Wu J 2018 Tribological properties of lubricating oil with micro/nano-scale WS_2 particles *J. Exp. Nanosci.* **13** 27–38
[77] Ilie F and Covaliu C 2016 Tribological properties of the lubricant containing titanium dioxide nanoparticles as an additive *Lubricants* **4** 12

[78] Zhao J, He Y, Wang Y, Wang W, Yan L and Luo J 2016 An investigation on the tribological properties of multilayer graphene and MoS_2 nanosheets as additives used in hydraulic applications *Tribol. Int.* **97** 14–20
[79] Charoo M S and Wani M F 2017 Tribological properties of h-BN nanoparticles as lubricant additive on cylinder liner and piston ring *Lubr. Sci.* **29** 241–54
[80] Wan Q, Jin Y, Sun P and Ding Y 2015 Tribological behaviour of a lubricant oil containing boron nitride nanoparticles *Proc. Eng.* **102** 1038–45
[81] Bao Y Y, Sun J L and Kong L H 2017 Tribological properties and lubricating mechanism of SiO_2 nanoparticles in water-based fluid *IOP Conf. Ser.: Mater. Sci. Eng.* **182** 012025
[82] Beheshti A, Huang Y, Ohno K, Blakey I and Stokes J R 2020 Improving tribological properties of oil-based lubricants using hybrid colloidal additives *Tribol. Int.* **144** 106130
[83] Wang X, Zhang Y, Yin Z, Su Y, Zhang Y and Cao J 2019 Experimental research on tribological properties of liquid phase exfoliated graphene as an additive in SAE 10W-30 lubricating oil *Tribol. Int.* **135** 29–37
[84] Sharif M Z, Azmi W H, Redhwan A A M and Mamat R 2016 Investigation of thermal conductivity and viscosity of Al_2O_3/PAG nanolubricant for application in automotive air conditioning system *Int. J. Refrig* **70** 93–102
[85] Asadi A, Asadi M, Rezaei M, Siahmargoi M and Asadi F 2016 The effect of temperature and solid concentration on dynamic viscosity of MWCNT/MgO (20–80)–SAE50 hybrid nano-lubricant and proposing a new correlation: an experimental study *Int. Commun. Heat Mass Transfer* **78** 48–53
[86] Jatti V S and Singh T P 2015 Copper oxide nano-particles as friction-reduction and anti-wear additives in lubricating oil *J. Mech. Sci. Technol.* **29** 793–8
[87] Suganthi K S, Manikandan S, Anusha N and Rajan K S 2015 Cerium oxide–ethylene glycol nanofluids with improved transport properties: preparation and elucidation of mechanism *J. Taiwan Inst. Chem. Eng.* **49** 183–91
[88] He Q, Li A, Wang Z, Zhang Y, Kong L and Yang K 2018 Tribological behavior of ZnO–Si_3N_4 nanoparticles-based lubricating grease *J. Exp. Nanosci.* **13** 231–44
[89] Ali M K A, Xianjun H, Mai L, Bicheng C, Turkson R F and Qingping C 2016 Reducing frictional power losses and improving the scuffing resistance in automotive engines using hybrid nanomaterials as nano-lubricant additives *Wear* **364** 270–81
[90] Ren B, Gao L, Li M, Zhang S and Ran X 2020 Tribological properties and anti-wear mechanism of ZnO@ graphene core-shell nanoparticles as lubricant additives *Tribol. Int.* **144** 106114
[91] Jia X, Huang J, Li Y, Yang J and Song H 2019 Monodisperse Cu nanoparticles@ MoS_2 nanosheets as a lubricant additive for improved tribological properties *Appl. Surf. Sci.* **494** 430–9
[92] Adams D R 2010 Tribological considerations in internal combustion engines *Tribology and Dynamics of Engine and Powertrain* (Amsterdam: Elsevier) pp 251–83
[93] Hasan M R 2017 Failure investigation report on different components of an automotive engine *Int. J. Mech. Eng. Appl.* **5** 47–51
[94] Katiyar J K, Bhattacharya S, Patel V K and Kumar V 2019 *Automotive Tribology* (Singapore: Springer)
[95] Gangopadhyay A 2017 A review of automotive engine friction reduction opportunities through technologies related to tribology *Trans. Indian Inst. Met.* **70** 527–35

[96] Radzevich S P 2016 *Dudley's Handbook of Practical Gear Design and Manufacture* (Boca Raton, FL: CRC Press)

[97] Anandika M N, Kahraman A and Talbot D 2019 An experimental investigation of the impact of random spacing errors on the dynamic transmission error of Spur Gear Pairs *Int. Design Engineering Technical Conf. and Computers and Information in Engineering Conf.* vol 59308 (New York: American Society of Mechanical Engineers), V010T11A024

[98] Brethee K F, Gu F and Ball A D 2016 Frictional effects on the dynamic responses of gear systems and the diagnostics of tooth breakages *Syst. Sci. Control Eng.* **4** 270–84

[99] Li Z and Mao K 2013 Frictional effects on gear tooth contact analysis *Adv. Tribol.* **2013** 181048

[100] Jelaska D T 2012 *Gears and Gear Drives* (New York: Wiley)

[101] Gupta K, Laubscher R F, Davim J P and Jain N K 2016 Recent developments in sustainable manufacturing of gears: a review *J. Clean. Prod.* **112** 3320–30

[102] Evans S M and Keogh P S 2019 Wear mechanisms in polyoxymethylene spur gears *Wear* **428** 356–65

[103] Diez-Ibarbia A, Fernandez-del-Rincon A, De-Juan A, Iglesias M, Garcia P and Viadero F 2017 Frictional power losses on spur gears with tip reliefs. The load sharing role *Mech. Mach. Theory* **112** 240–54

[104] Garcia J G, Misra R D K and Gonzalez E 2020 Fatigue failure of gears and bearings during processing of rebar steels *J. Fail. Anal. Prev.* **20** 208–17

[105] Schultheiss H, Tobie T and Stahl K 2018 The wear behavior of small module gears lubricated with NLGI 1 grease *Tribol. Trans.* **61** 229–37

[106] Morales-Espejel G E and Gabelli A 2018 A model for gear life with surface and subsurface survival: tribological effects *Wear* **404** 133–42

[107] Rincon A F D, Viadero F, Iglesias M, de-Juan A, Garcia P and Sancibrian R 2012 Effect of cracks and pitting defects on gear meshing *Proc. Inst. Mech. Eng.* C **226** 2805–15

[108] Fabre A, Evans H P, Barrallier L, Sharif K J and Desvignes M 2013 Prediction of microgeometrical influences on micropitting fatigue damage on 32CrMoV13 steel *Tribol. Int.* **59** 129–40

[109] Raadnui S 2019 Spur gear wear analysis as applied for tribological based predictive maintenance diagnostics *Wear* **426** 1748–60

[110] Dempsey P J 2000 A comparison of vibration and oil debris gear damage detection methods applied to pitting damage *NASA Technical Report* E-12423

[111] Patil M, Ramkumar P and Shankar K 2017 Multi-objective optimization of spur gearbox with inclusion of tribological aspects *J. Frict. Wear* **38** 430–6

[112] Patil M, Ramkumar P and Shankar K 2019 Multi-objective optimization of the two-stage helical gearbox with tribological constraints *Mech. Mach. Theory* **138** 38–57

[113] Onishchenko V 2015 Investigation of tooth wears from scuffing of heavy duty machine spur gears *Mech. Mach. Theory* **83** 38–55

[114] Jiang Y, Choudhury D, Brownell M, Nair A, Goss J A and Zou M 2019 The effects of annealing conditions on the wear of PDA/PTFE coatings *Appl. Surf. Sci.* **481** 723–35

[115] Liu H, Liu H, Zhu C, Wei P and Tang J 2019 Tribological behavior of coated spur gear pairs with tooth surface roughness *Friction* **7** 117–28

[116] Kameyama Y, Takahashi R, Owaku Y, Sato H and Shimpo R 2014 Effect of shot particle conditions on the transfer of copper induced by fine particle peening *Proc. ICSP* 12 450–5

[117] Zhang Y L, Lai F Q, Qu S G, Liu H P, Jia D S and Du S F 2019 Effect of ultrasonic surface rolling on microstructure and rolling contact fatigue behavior of 17Cr2Ni2MoVNb steel *Surf. Coat. Technol.* **366** 321–30

[118] Rajendhran N, Palanisamy S, Shyma A P and Venkatachalam R 2018 Enhancing the thermophysical and tribological performance of gear oil using Ni-promoted ultrathin MoS_2 nanocomposites *Tribol. Int.* **124** 156–68

[119] Jendrzej S, Gondecki L, Debus J, Moldenhauer H, Tenberge P, Barcikowski S and Gökce B 2019 Tribological properties of laser-generated hard ceramic particles in a gear drive contact *Appl. Surf. Sci.* **467** 811–8

[120] Vandoni L, Demir A, Previtali B, Lecis N and Ugues D 2012 Wear behavior of fiber laser textured TiN coatings in a heavy loaded sliding regime *Materials* **5** 2360–82

[121] Priyadharshan S J, Srinivas V, Chebattina K R R and Vandana V 2019 Tribological characteristics of epoxide based bio lubricant–commercial gear oil blends *IOP Conf. Ser.: Mater. Sci. Eng.* **653** 012041

[122] Doll K M, Heise G L, Myslinska M and Sharma B K 2012 Formulation of a biobased gear oil utilizing boron technology *Int. Joint Tribology Conf.* vol 45080 (New York: American Society of Mechanical Engineers), 351–3

[123] Cigno E, Magagnoli C, Pierce M S and Iglesias P 2017 Lubricating ability of two phosphonium-based ionic liquids as additives of a bio-oil for use in wind turbines gearboxes *Wear* **376** 756–65

[124] Saravanakumar N, Jothi Saravanan M L, Barathkumar K E, Gokula Kannan K and Karthikeyan R 2019 Development and testing of nano particulate lubricant for worm gear application *J. Mech. Sci. Technol.* **33** 1785–91

[125] Ahmad A, Khurram M, Mufti R A and Umar M 2020 Simulating cam and follower wear in valve train *Tribol.-Mater., Surf. Interfaces* **14** 59–65

[126] Arshad W, Hanif M A, Bhutta M U, Mufti R A, Shah S R, Abdullah M U and Najeeb M H 2017 Technique developed to study camshaft and tappet wear on real production engine *Ind. Lubricat. Tribol.* **69** 174–81

[127] Shahriman A B, Syafiq A K M, Hashim M S M, Hazry D, Razlan Z M, Wan K and Azizan A 2017 Improvement of cam performance curve using B-Spline curve. *J. Phys.: Conf. Ser.* **908** 012043

[128] Alakhramsing S S, de Rooij M, Schipper D J and van Drogen M 2018 Lubrication and frictional analysis of cam–roller follower mechanisms *Proc. Inst. Mech. Eng. J.* **232** 347–63

[129] Al-Jeboori Y, Kosarieh S, Ofune M, Neville A and Morina A 2018 The effect of clearance between tappet insert and camlobe on the tribological and tribochemical performance of cam/follower surfaces *Tribol. Int.* **149** 105409

[130] Jamali H U, Al-Hamood A, Abdullah O I, Senatore A and Schlattmann J 2019 Lubrication analyses of cam and flat-faced follower *Lubricants* **7** 31

[131] Alakhramsing S S, de Rooij M, Schipper D J and van Drogen M 2018 Lubrication and frictional analysis of cam–roller follower mechanisms *Proc. Inst. Mech. Eng. J.* **232** 347–63

[132] Hoag K and Dondlinger B 2016 Cylinder head layout design *Vehicular Engine Design* (Vienna: Springer) pp 147–76

[133] Godiño J A V, García M T, Aguilar F J J E and Guerrero D P 2019 Failure analysis of an overhead valve train system in urban buses *Eng. Fail. Anal.* **96** 455–67

[134] Tanaka K and Arase K 2019 *US Patent* 10,253,658

[135] Basu A, Gokhale N, Aghav Y and Kumar M N 2020 Design of valve train for heavy duty application *Design and Development of Heavy Duty Diesel Engines* (Singapore: Springer) pp 601–36
[136] Kono S 2019 *US Patent* 10,227,902
[137] Zhao Y, Zhang K and Di J 2018 Structural design of two-cylinder single overhead camshaft *IOP Conf. Ser.: Mater. Sci. Eng.* **301** 012131
[138] Lee J, Rydzewski J, Gardner D and Bowyer S 2019 The new Blackwing V8 engine in the Cadillac CT6 V *MTZ Worldwide* **80** 88–94
[139] Lawes S D A, Hainsworth S V and Fitzpatrick M E 2010 Impact wear testing of diamond-like carbon films for engine valve-tappet surfaces *Wear* **268** 1303–8
[140] Jana P 2018 Automation in sewing technology *Automation in Garment Manufacturing* (Cambridge: Woodhead Publishing) pp 199–236
[141] Yu C, Meng X and Xie Y 2017 Numerical simulation of the effects of coating on thermal elastohydrodynamic lubrication in cam/tappet contact *Proc. Inst. Mech. Eng. J.* **231** 221–39
[142] Wu W, Wang J and Venner C H 2016 Thermal elastohydrodynamic lubrication of an optimized cam–tappet pair in smooth contact *J. Tribol.* **138** 021501
[143] Xin Q 2011 *Diesel Engine System Design* (Amsterdam: Elsevier)
[144] Chen F, Chang Y, Xie Z, Zhang K, Sun T, Li X and Huang Y 2018 Simulation and experimental research of hydraulic pressure and intake valve lift on a fully hydraulic variable valve system for a spark-ignition engine *Adv. Mech. Eng.* **10** 1687814018773156
[145] Umar M, Mufti R A and Khurram M 2019 Comparison of flash temperature at cam/tappet interface using different theoretical models and its effect on friction torque *Tribol. Trans.* **63** 425–38
[146] Gangopadhyay A and McWatt D G 2006 The effect of novel surface textures on tappet shims on valvetrain friction and wear *Int. Joint Tribology Conf.* vol 42592 533–42
[147] Tatarko P and Kašiarová M 2015 Silicon nitride based ceramic composites toughened by rare-earth oxide additives *Toughening Mechanisms in Composite Materials* (Cambridge: Woodhead Publishing) pp 35–71
[148] Guo X, Cai Y, Lai F, Wang J, Li Z and Qu S 2019 Study on fretting wear properties of high speed diesel engine roller tappet material Cr12W steel *IOP Conf. Ser.: Mater. Sci. Eng.* **569** 022022
[149] Gangopadhyay A, McWatt D G, Zdrodowski R J, Simko S J, Peczonczyk S L, Cuthbert J and Hock E D 2018 Valvetrain friction and wear performance of polyalkylene glycol engine oils *Tribol. Trans.* **61** 133–43
[150] Singh P and Goel V 2018 Effect of bio-lubricant on wear characteristics of cylinder liner-piston ring and cam-tappet combination in simulated environment *Fuel* **233** 677–84
[151] Zahid R, Hassan M H, Alabdulkarem A, Varman M, Kalam M A, Mufti R A and Abdullah U 2018 Tribological characteristics comparison of formulated palm trimethylol-propane ester and polyalphaolefin for cam/tappet interface of direct acting valve train system *Ind. Lubricat. Tribol.* **70** 888–901
[152] Singh P and Goel V 2018 Effect of bio-lubricant on wear characteristics of cylinder liner-piston ring and cam-tappet combination in simulated environment *Fuel* **233** 677–84
[153] Ruddy B L, Dowson D and Economou P N 2017 A review of studies of piston ring lubrication *Proc. 9th Leeds-Lyon Symp. on Tribology* 109–21
[154] Rahmani R, Rahnejat H, Fitzsimons B and Dowson D 2017 The effect of cylinder liner operating temperature on frictional loss and engine emissions in piston ring conjunction *Appl. Energy* **191** 568–81

[155] Söderfjäll M, Almqvist A and Larsson R 2016 Component test for simulation of piston ring–cylinder liner friction at realistic speeds *Tribol. Int.* **104** 57–63

[156] Fischer M, Lammers R, Bauer C and Hoppe S 2018 *US Patent* 10,000,840

[157] Zhang Z, Liu J and Xie Y 2016 Design approach for optimization of a piston ring profile considering mixed lubrication *Friction* **4** 335–46

[158] Bifeng Y, Dashu G, Shao S, Bo X and Hekun J 2018 Research on the profile design of surface texture in piston ring of internal combustion engine *J. Tribol.* **140** 061701

[159] Mishra P C 2014 A review of piston compression ring tribology *Tribol. Ind.* **36** 269–80

[160] Delprete C and Razavykia A 2018 Piston ring–liner lubrication and tribological performance evaluation: a review *Proc. Inst. Mech. Eng. J.* **232** 193–209

[161] Dimkovski Z, Tomanik E and Profito F 2016 Influence of measurement and filtering type on friction predictions between cylinder liner and oil control ring *Tribol. Int.* **100** 7–17

[162] Degawa S, Kobayashi W, Kikuhara K, Nishibe K and Ito A 2019 Numerical study on pressure generation mechanism of oil film under oil control ring *SAE Technical Paper* 2019-1-2361

[163] Fang T and Tian T 2016 Oil transport from scraper ring step to liner at low engine speeds and effect of dimensions of scraper ring step *SAE Int. J. Fuels Lubr.* **9** 48–58

[164] Zhang Y, Zhang X, Wu T and Xie Y B 2016 Effects of surface texturing on the tribological behavior of piston rings under lubricated conditions *Ind. Lubricat. Tribol.* **68** 158–69

[165] Ma W, Biboulet N and Lubrecht A A 2018 Performance evolution of a worn piston ring *Tribol. Int.* **126** 317–23

[166] Wong V W and Tung S C 2016 Overview of automotive engine friction and reduction trends—effects of surface, material, and lubricant-additive technologies *Friction* **4** 1–28

[167] Zabala B, Igartua A, Fernández X, Priestner C, Ofner H, Knaus O and Nevshupa R 2017 Friction and wear of a piston ring/cylinder liner at the top dead centre: experimental study and modelling *Tribol. Int.* **106** 23–33

[168] Mohamed E S 2018 Performance analysis and condition monitoring of ICE piston-ring based on combustion and thermal characteristics *Appl. Therm. Eng.* **132** 824–40

[169] Cheng J, Meng X, Xie Y and Kong X 2016 Blow-by and tribological performance of piston ring pack during cold start and warm idle operations *Sci. China Technol. Sci.* **59** 1085–99

[170] Moosavian A, Najafi G, Ghobadian B and Mirsalim M 2017 The effect of piston scratching fault on the vibration behavior of an IC engine *Appl. Acoust.* **126** 91–100

[171] Notay R S, Priest M and Fox M F 2019 The influence of lubricant degradation on measured piston ring film thickness in a fired gasoline reciprocating engine *Tribol. Int.* **129** 112–23

[172] Tomanik E, Profito F, Sheets B and Souza R 2018 Combined lubricant–surface system approach for potential passenger car CO_2 reduction on piston-ring-cylinder bore assembly *Tribol. Int.* **149** 105514

[173] Grabon W, Pawlus P, Wos S, Koszela W and Wieczorowski M 2017 Effects of honed cylinder liner surface texture on tribological properties of piston ring-liner assembly in short time tests *Tribol. Int.* **113** 137–48

[174] Söderfjäll M, Herbst H M, Larsson R and Almqvist A 2017 Influence on friction from piston ring design, cylinder liner roughness and lubricant properties *Tribol. Int.* **116** 272–84

[175] Grabon W, Pawlus P, Wos S, Koszela W and Wieczorowski M 2018 Evolutions of cylinder liner surface texture and tribological performance of piston ring-liner assembly *Tribol. Int.* **127** 545–56

[176] Shen C and Khonsari M M 2016 The effect of laser machined pockets on the lubrication of piston ring prototypes *Tribol. Int.* **101** 273–83
[177] Wan S, Li D, Tieu A K and Zhang B 2017 Comparison of the scuffing behaviour and wear resistance of candidate engineered coatings for automotive piston rings *Tribol. Int.* **106** 10–22
[178] Kumar V, Sinha S K and Agarwal A K 2019 Tribological studies of an internal combustion engine *Advanced Engine Diagnostics* (Singapore: Springer) pp 237–53
[179] Peng Y, Xu Y, Geng J, Dearn K D and Hu X 2017 Tribological assessment of coated piston ring-cylinder liner contacts under bio-oil lubricated conditions *Tribol. Int.* **107** 283–93
[180] Arumugam S and Sriram G 2012 Effect of bio-lubricant and biodiesel-contaminated lubricant on tribological behavior of cylinder liner–piston ring combination *Tribol. Trans.* **55** 438–45
[181] Arumugam S and Sriram G 2013 Synthesis and characterisation of rapeseed oil bio-lubricant–its effect on wear and frictional behaviour of piston ring–cylinder liner combination *Proc. Inst. Mech. Eng. J.* **227** 3–15
[182] Gul M, Zulkifli N W M, Masjuki H H, Kalam M A, Mujtaba M A, Harith M H and Farooq A B 2020 Effect of TMP-based-cottonseed oil-biolubricant blends on tribological behavior of cylinder liner-piston ring combinations *Fuel* **278** 118242
[183] Peng Y, Xu Y, Geng J, Dearn K D and Hu X 2017 Tribological assessment of coated piston ring-cylinder liner contacts under bio-oil lubricated conditions *Tribol. Int.* **107** 283–93
[184] Arumugam S and Sriram G 2014 Synthesis and characterization of rapeseed oil bio-lubricant dispersed with nano copper oxide: its effect on wear and frictional behavior of piston ring–cylinder liner combination *Proc. Inst. Mech. Eng. J.* **228** 1308–18
[185] Cheah M Y, Ong H C, Zulkifli N W M, Masjuki H H and Salleh A 2020 Physicochemical and tribological properties of microalgae oil as biolubricant for hydrogen-powered engine *Int. J. Hydrogen Energy* **45** 22364–81

Chapter 5

Machine learning to develop lubrication strategies

Ashish K Kasar, Parker Hill, Soumya Sikdar and Pradeep L Menezes

Machine learning (ML) in tribology is of interest to engineers and scientists for its ability to predict certain tribological properties. Given the rapid pace of advancement in computing power, it is a given that ML will become more important. Specifically, the design and further optimization of artificial neural networks (ANNs), which draw inspiration from biological neural networks, is necessary moving forward. This paper will categorize and expound on various ML techniques as well as demonstrate the successes that have occurred using these techniques in experimentation. Regarding the future of ML in the field of tribology, this review will utilize the effectiveness of the methods as a concrete reason to further integrate these methods in laboratories around the globe. In particular, the focus will be on the advancement of artificial intelligence (AI) as the key technology to improve the efficiency and effectiveness of ML techniques.

5.1 Introduction

5.1.1 History of machine learning

The roots of ML are found in the early 1940s. Neurophysiologist Warren McCulloch and mathematician Walter Pitts wrote a paper on the nature of neurons as being 'on or off' information storage devices. They modeled nets of neurons as being in an electric circuit, with synapses being the triggers for the circuits. This is the first documented instance where it is proposed that organic neural behavior could be used to model mathematics and have applications in fields removed from biology [1]. A few years later in 1958, the first artificial neural network was designed by Frank Rosenblatt, a psychologist. Rosenblatt named the network Perceptron and its primary function was to model how the human brain would receive and process visual data into object recognition [2]. Perceptron was built on the premise that biological neurons and the binary code upon which computers are built are

doi:10.1088/978-0-7503-3435-8ch5

functionally similar. This concept is important to the fundamental idea of an ANN. An ANN is an algorithm that is directly modeled on the idea of organic neural networks. Each 'neuron' within the ANN, referred to as a threshold unit, receives data from the network, weights the data according to some standard, and then sums the data. The output is a value between 0 and 1 [3]. The Perceptron model for an ANN contained only one of these threshold units but was still considered revolutionary for its time. Single-layer ANNs will have only one layer of these threshold units between input and output layers. A multi-layered network could theoretically have an infinite number of these layers, but typically only has two or three.

Backpropagation, or backward propagation of errors, is an algorithm that allows an ANN to learn from past errors and more accurately analyze future data. First developed in the 1970s, it became significant following the writing of, 'Learning representations by back-propagating errors' by Rumelhart, Hinton, and Williams in 1986 [4]. Before the usage of this technique, ANN had struggled with finding a way for a network to learn good 'internal representations'. An internal representation as shown in figure 5.1, is a member of the 'hidden units' layer. These layer(s) consist of more weights and values that further allow the network to be more finely tuned, producing desired outputs from a larger range of data with less error [5]. Prior to this discovery, neural networks consisted of only two layers, input and output. With the usage of backpropagation, neural networks and thus ML were able to be applied to a more diverse range of problems.

Following the discovery of backpropagation as a critical ML algorithm, boosting became relevant in 1990. First mentioned in 'The strength of weak learnability', by Robert Schapire, boosting refers to the ability of an algorithm to reduce supervised learning bias, as well as turn weak learners into strong ones [6]. Weak learners are considered any algorithm that, when given inputs, produces outputs that are only slightly more accurate than random guessing would yield. In contrast, a strong

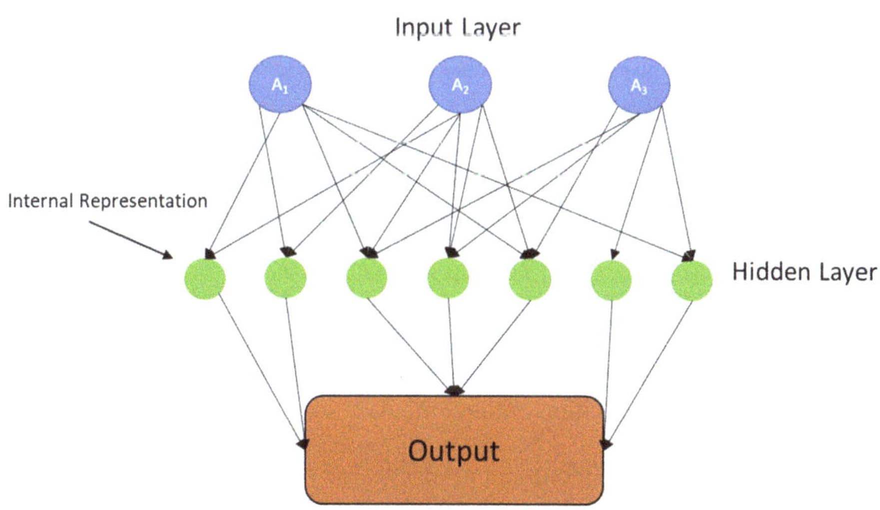

Figure 5.1. Representation of ANN.

learner algorithm will output data with high confidence and low error for all inputs which the algorithm was built to handle. The process initializes by grouping a large number of weak learners. Each algorithm individually analyzes desired data one at a time, assigning weights to individual data points. Each subsequent weak learner after the first will assign more weight to data points misclassified by its predecessor and less weight to points accurately classified. This process is repeated until an algorithm is of the desired accuracy [7]. Boosting is so important because it allows for algorithms that would otherwise have no usage of value.

Each of the previously mentioned points in history was significant to the growth of ML and its subsequent growing reputation as a highly useful technique. Moving into the 21st century, the usage of ML has grown exponentially, including in the field of tribology [8] for designing surface texture, mixed-lubricants, and composite materials.

5.1.2 Artificial intelligence versus machine learning

When discussing ML, it is important to note the distinction between ML and artificial intelligence as there is widespread confusion on the topic. Though similar in the sense that both methodologies focus on the use of computing power for problem-solving, they are distinct in definition and often operation. Essentially the term artificial intelligence can be applied to any computer program that has human-like behavior. However, if the algorithm does not inherently learn anything from input data, it cannot be classified as ML [3]. Thus, all ML techniques are considered AI, but all AI techniques cannot be considered ML.

5.2 General ML tasks

ML as a field is extremely broad, with usage in the worlds of engineering, medicine, finance, and humanities. At its core, ML is translating human learning techniques into computational algorithms. The algorithms rely heavily on mathematic techniques, specifically statistical methods. The 'learning' side of ML refers to the algorithm's eventual ability to draw from experience, just as a human learner would, to solve a variety of potential tasks. The output of ML is a prediction, wherein the machine draws on all previously learned experiences (inputted by the user) and draws a conclusion using a programmed algorithm. Thus, it is of the utmost importance that the inputted 'experience' is as accurate as possible to mitigate prediction error. This characteristic of ML leads to several key observations. Firstly, because successful ML does rely on a pool of experience, ML will only become more and more accurate in predictive reasoning as the technique has an increasing amount of data to rely upon. Regardless of the application, this technique only increases in effectiveness over time.

Though inherently involving prediction, ML tasks can still be classified off desired output qualification. The classification consists of defining a label or category for individual data points in a set [9]. For example, given a set of images consisting of the surface of various metals, ML would need to categorize each image as being steel, aluminum, copper, etc. Ranking involves a system ordering data sets against a specific requirement [9]. Tribologically, this could involve ranking profilometer images from lowest to highest surface roughness. In addition, clustering technique attempts to analyze and display trends between large data points. Often used in larger data sets, homogenous clusters, wherein some overarching trait is shared, are displayed to the user for easy analysis [9]. While further ML tasks do exist, the previously mentioned tasks are among the most commonly used.

5.3 Support vector machine

Essentially, a support vector machine (SVM) is a classifier algorithm that uses example utilization to continuously learn [10]. Four basic concepts are essential to understanding how an SVM operates: separating hyperplane, maximum-margin hyperplane, soft margin, and the kernel function. The separating hyperplane (represented in figure 5.2) is a 3-D representation of a line on a 2-D graph that separates clusters of data. A maximum-margin hyperplane adds the stipulation that the separating hyperplane must also be in a position to maximize the magnitude of a vector taken from the hyperplane to either set of data. The previous two concepts assume that a line can be placed directly between the two theoretical sets of data. Unfortunately, this assumption is often not realistic, thus there is usage of a soft margin. A soft margin enables an SVM algorithm to allow data points to push through the hyperplane without affecting the final result. However, too many of these allowed points introduce an uncontrollable amount of error into the SVM prediction, thus, a user-defined parameter is necessary that allows for the specification of the number of points that may exhibit this behavior. The kernel function

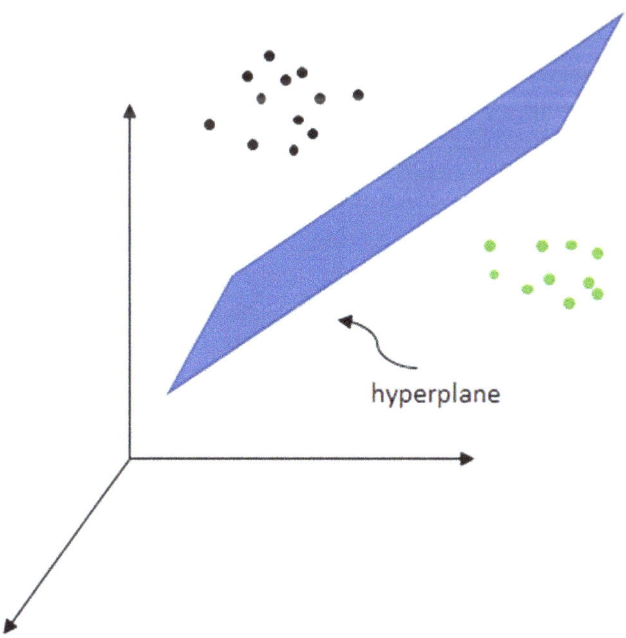

Figure 5.2. Representation of hyperplane in 3-D space.

allows an SVM to project data into an additional dimension, enabling data that could not be classified with a simple hyperplane or soft margin to be sorted more accurately [11]. The SVM algorithm is highly useful and has had many usages in the field of ML.

5.4 Genetic algorithm

The genetic algorithm is a highly useful ML optimization algorithm that closely models the process of natural selection. The algorithm works by taking a random selection of data points and then giving the opportunity to those data points that are desirable to the project outcome 'reproduce'. The initial model that most research and usage of this technique comes from a model first introduced by John Holland in 1975 [12]. The genetic algorithm is most often used in conjunction with some other ML technique, such as the ANN or the SVM.

5.5 Applications in tribology

As demonstrated, ML techniques can be applied to a vast range of topics. When utilized in tribology, ML has been successful in the design of lubricants and alloys, monitoring internal wear conditions of complex systems and predicting tribological properties.

5.5.1 Formulation and usage of lubricants

ML techniques have been effectively adapted to build materials customized to specific goals. For example, Wan *et al* [13] discovered new ester lubricant formulations using applied ML techniques. Ester lubricants are prized for their low volatility and high lubricating properties, and are widely used in industrial applications. These synthetic lubricants are produced by the esterification of acids and alcohols. The molecular structure of these ester lubricants is responsible for their anti-frictional properties. Thus, the researchers felt the need to build a system that correctly modeled this relationship to produce desirable ester lubricants. Several ML algorithms were applied for this purpose, including least squares-support vector regression (LS-SVR) and multiple stepwise regression (MSR). Molecular properties, such as relative van der Waal's volume, were used as inputs for this experiment, with a coeficient of friction (COF) value being the output. Over 50 different lubricant molecules were tested, with the mean relative error (MRE) of the difference between the expected value and ML predicted value for the training set being 2.92%. The MRE of the test set was found to be 6.03%. These numbers reflect the success of the ML techniques in predicting the COF of many substances based on molecular descriptors. The authors concluded that the anti-wear performance of lubricants can be optimized by using ML techniques.

To support the creation of green lubricants, Bhaumik *et al* [14] designed a new modified bio-lubricant using several custom ANNs and a genetic algorithm. Firstly, the most important friction modifiers were determined using ANN. These were found to be load, speed, graphite concentration, graphene concentration, multi-walled carbon nanotube concentration, and zinc oxide concentration. With the objective of minimizing the COF, and taking the output information generated by the ML techniques, lubricant formulation was completed. The lubricant was then tested using pin-on-disc and four-ball tribometer setups. The resulting wear tracks were imaged with an optical microscope, scanning electron microscope, and micro Raman spectrograph. The new bio-lubricant, being modified vegetable oil, and compared to commercial mineral oil, was found to reduce the COF by 50%. Wear scar diameter was reduced by 87%. It was concluded that the ML approach was effective and would be beneficial to future lubricant formulation projects.

ANN techniques have been used widely to design vegetable-based oil lubricants and predict their performance. For example, Edla *et al* [15] investigated trans esterified vegetable oil blends of rice bran oil and jatropha oil for formulating metal-working fluids. ML techniques of linear and polynomial regression were used to predict the cutting forces, which were also experimentally verified by conducting experiments using a lathe machine equipped with a dynamometer. The polynomial regression showed an accuracy of 99.1%, which was higher than the accuracy in the case of linear regression (98.88%). Similarly, Bharadwaj *et al* [16] employed an ANN model to investigate and predict surface roughness and COF values for different lubricants such as mineral oil and modified vegetable oils based on the machining conditions of the surfaces. Therefore, the input values were the properties of oil and machining conditions to predict the value of average surface roughness (Ra) and COF. The lubricant properties include kinematic viscosity, viscosity index, saponification value, acid value, flash point, and pour point, whereas the inputs for machining conditions were operational variables of sliding, cutting, post sliding, total time, and RPM. It was observed that the ANN model could predict surface roughness values more accurately compared to the coefficient of friction values. This study highlighted that the well-designed ANN model had an accuracy of more than 83.6 and 99.8% for the COF and surface roughness compared to the tedious tribo-testing experiments. Bhaumik *et al* [17] developed an ANN model to predict the COF and wear properties of various vegetable oil-based lubricants incorporated with nano friction modifier additives. The studied vegetable oils were coconut oil, castor oil, and palm oil with graphene and multi-walled carbon nanotubes as nano friction modifier additives. The lubricants were also investigated experimentally using a pin-on-disk and four-ball tester to compare the predicted results with the experimental results. In this study, using the ANN model, the authors designed a new lubricant by optimizing 11–13 variables, such as load, sliding speed, and temperature. The newly designed lubricant was observed to exhibit enhanced tribological properties. Humelnicu *et al* [18] investigated using the ANN technique the tribological behavior of sunflower oil and rapeseed oil added to diesel fuel to form bio-diesel (figure 5.3). Using this technique, the authors optimized the amount of vegetable oil in the diesel to achieve superior lubrication performance. The results stated that lubricant mixture with 4% sunflower oil in the diesel can yield the lowest COF (~0.001 56), whereas a higher concentration of rapeseed oil (19%) was required to achieve the lowest COF (approximately 0.011) in the lubricant mixture of rapeseed oil and diesel. This study using the ANN algorithm concluded that the lubricant mixture with sunflower oil provided the best performance, which was correlated with experimental studies.

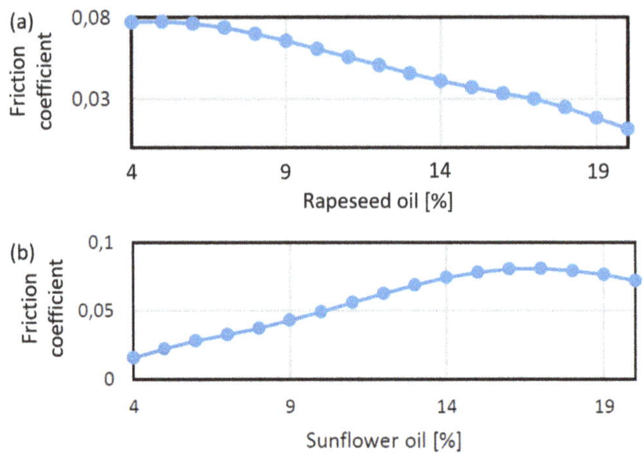

Figure 5.3. Influence of on COF values of diesel incorporated with different vol % of (a) rapeseed oil, and (b) sunflower oil. Reproduced from [18] CC BY 4.0.

Mujtaba *et al* [19] demonstrated the prediction of bio-diesel production using a response surface methodology and extreme ML techniques. The bio-diesel was synthesized from palm–sesame oil using ultrasound-assisted transesterification process. Factors such as mean absolute percentage error, coefficient of determination, mean square error, root mean square error, and standard error of prediction were calculated to assess the above-mentioned ML techniques. It was observed from this study that the extreme learning machine technique provided a better result compared to the response surface methodology technique.

Although literature specifically related to material design is relatively lacking, through the existing references it is clear to see the power of ML techniques in material design. Because ML allowed for massive amounts of data to be sorted relatively quickly, relationships between existing inputs and desired outputs are efficiently found. This enables researchers to selectively target the material design parameters needed and formulate without significant trial and error.

The application of ML is not limited to liquid lubricant design, it has also been applied to design composite materials for tribological applications. For instance, Dey *et al* [20] conducted experimentation to design age-hardened aluminum alloys of a specific desired hardness using ANN models as a basis for the multi-objective genetic algorithm (MOGA). It was concluded that the ANN models were practical for designing alloys. Vinoth and Datta designed ultra-high molecular weight polyethylene composites using both ANN and genetic algorithm (GA) techniques [21]. Correlating such values as nanoparticle type, volume fraction, and size with the resulting material properties can be difficult. Thus, the ML techniques replaced this complex correlation operation and used weight percent and particle size as inputs into the ML, with mechanical properties being the outputs. The ANN was a feed-forward network using a scaled conjugate gradient backpropagation algorithm for training. The genetic algorithm, being comprised of selection, crossover, and mutation operations, repeatedly modified a selected sample size of individual solutions. The ML operations used were successful in revealing the function of nanoparticle variables in creating mechanical properties, such as elastic modulus and material hardness. Moreover, the ML techniques enabled the researchers to enhance said properties significantly.

5.5.2 Wear condition monitoring

The earliest recorded usage of ML in tribological research occurs in the paper 'Characterization of wear particles and their relations with sliding conditions', written by Umeda *et al* in 1997. In this work, the authors utilized two ANNs to make predictions based on wear particle images from sliding pin-on-disk tests [22]. Qualifiers used to make classifications were representative diameter, roundness, elongation, and reflectivity. With these qualifiers, the algorithm successfully predicted the specific pin-on-disk sliding conditions used for any given image and the resulting coefficient of friction value. Given that the algorithm was successful in prediction, this work suggests that the same technique could also be used in wear condition monitoring in addition to lubricant design and predicting their performance.

Over two decades later, the previous authors have been proven correct. Konig *et al* [23] utilized a multi-layer classification network to predict critical operating conditions of sliding bearing systems used in wind turbine planetary gearboxes. According to the authors, turbine gearbox design is bottlenecked by the reliability of a sliding bearing when operating within varying frictional conditions. Normal turbine operation can cause the bearing lubricant to become contaminated with wear particles from gears within the box as well as sand used in the manufacturing process. Particle-contaminated lubricant as well as an inadequate quantity of lubrication, causes wear conditions to form that can cause catastrophic failure to the system. To mitigate this, acoustic emission (AE) has become a potentially useful wear-monitoring technique. AE monitors a given system for ultrasonic stress waves generated by material deformation in localized locations within said system. The authors monitored various shaft conditions such as particle-contaminate lubrication, oil-starvation, and run-in conditions using an in-house test rig. It was found that the ANN given the AE signals as input was able to classify critical wear conditions for the oil-starvation and run-in conditions with an accuracy of 82.5%, while in the case of particle contamination the accuracy was 100%. It was proposed that the lower accuracy in the case of oil-starvation and run-in conditions was due to the similarity of the two processes.

Journal bearings in turbofan aircraft engines experience the same problems encountered within wind turbine gearboxes. Mokhtari *et al* [24] utilized ML techniques to monitor and classify AE signals generated during turbofan operation. Two experimental test rigs were built consisting of oil-lubricated journal bearings and driven shafts. Three friction states were induced throughout the experimentation: dry, a transition state between dry and fluid, and fully flooded with fluid. Utilizing an SVM classifier, the AE signals generated were acquired, processed, and classified into one of the previously mentioned categories. When experimentation was finished, it was found that the algorithm had operated with a detection accuracy rate of 96.7%, even with different oil viscosities being used in operation. In this case, the SVM allowed the researchers to quickly classify various condition states and improve operating procedures. Kankar *et al* [25] applied both ANN and SVM to detect surface faults on operating ball bearings. Using an experimental test rig in combination with accelerometers to acquire vibrational signals, testing was conducted. It was found that the SVM performed better at classifying signals as surface faults. It was noted, however, that both techniques performed excellently and proved that ML techniques can be applied as early-warning systems to signal when maintenance is needed, preventing expensive, dangerous catastrophic failure.

Similar to wear conditioning monitoring, ML techniques have also been applied to surface coatings. Surface coatings are applied to various materials to improve the wear performance at the interface, prolonging machine life. The process of testing surface coatings can be time-consuming and, due to the number of experiments involved, intrinsically costly. Thus, ML techniques are being applied with great advantages to reduce the time of experimentation. Recently, Altay *et al* [26] employed linear regression (LR), SVM, and Gaussian process regression (GPR) algorithms to predict wear loss of iron-based alloys which had been welded to test material, thus improving the wear resistance of the sample. Tribo-testing was conducted on the sample using a block-on-disc adhesive tester and conducted at progressively increasing loads. Input parameters, such as powder composition, the average microhardness of coating layers, the load applied, and sliding distance, were used to train the algorithms. The researchers found that the LR algorithm was the least accurate, with an R-squared success rate of 0.93. The SVM and GPR methods yielded calculated accuracies of 0.96. Although SVM and GPR were similar in accuracy, it was proposed that future processing would benefit from an increased focus on the GPR technique due to its better performance in categories used to calculate the R-squared value. Through using these ML techniques, the researchers were able to reduce the number of experiments required for wear condition monitoring, which in turn reduced the necessary project budget.

Increasing reliability in large engine systems can be effectively targeted by improving the performance of tribological interactions within these systems. Xu *et al* [27] have sought to improve wear fault prediction within large maritime engine systems through the fusion of several data-driven ML and AI techniques. These techniques included ANNs, belief rule-based interference (BRB), and evidential reasoning (ER). In this process, information is taken from the monitoring of engine oil (wear particles resulting from different wear mechanisms, oxides evidencing corrosion) were used as input to the algorithms. Typically, the analysis of such information by engineers is difficult due to unpredictability in the relationship between wear fault values and actual fault states. However, the ML techniques employed here enabled the researchers to accurately predict wear fault states from information that would otherwise be vague and complex to a human analyzer. ER rule allows for independent evidence to be fused into a probabilistic reasoning process that considers both reliability and the important weight of said evidence. Indeed, the fused diagnostic model created by Xu *et al* [27] utilized the ER rule to integrate three separate single-data-driven models (ANN, BRB, ER). It was concluded that using a fused diagnostic model resulted in a system that was significantly more robust and accurate, increasing overall accuracy by as much as 12.8%. It is noted, however, that the performance of any single model had a significant effect on the accuracy of the fused model as a whole. Thus, any inaccuracy in any one of the fused techniques led to a major skew in the diagnostic data.

5.5.3 Value prediction

Aside from wear conditioning monitoring, ML techniques have also been applied with great success to material property prediction tasks. Borjali *et al* [28] used ML techniques in conjunction with pin-on-disk experimentation to optimize research on polyethylene hip implant research. Polyethylene wear occurs during normal use of implants made from these materials, and the introduction of this wear debris into the body may cause osteolysis. Osteolysis is progressive destruction of bone tissue, which left untreated, leads to improper fit of the implant and reduction of vital mechanical properties. The researchers utilized three ML sub-fields in conjunction with the traditional pin-on-disc experimentation. The first approach utilized interpretable model-based methods, such as linear regression and gradient boosting. These techniques allow a model to be trained based on data from experimentation and then make predictions. The second approach involved non-interpretable model-based methods, specifically ANN and SVM algorithms. The third approach was instance-based methods. This method, referred to as the *k*-nearest neighbor (KNN) method, is useful for handling complicated data sets and predicts the wear rate of most similar instances in the experimentation data sets. It was found that the KNN technique resulted in the lowest prediction error and theorized that due to the usability of the ML technique, it could result in less need for physical experimentation in the future.

Copper, well known for its excellent conductive properties, unfortunately must be strengthened and hardened when the application requires it to be non-ductile. This can be accomplished through the even dispersion of ceramic particles throughout the copper matrix. Thankachan *et al* [29] introduced a mixture of aluminum nitride (AlN) and boron nitride (BN) particles into the copper matrix through friction-stir processing. The resulting composite material was then analyzed using traditional methods to determine if there had been increased resistance to friction from the material. Using the data acquired during traditional experimentation, an ANN was then trained to predict the future wear rates of the resultant copper–ceramic composite. Hayajneh *et al* [30] also worked to optimize composite materials and utilized an ANN to predict the tribological behavior of the new materials. Aluminum–copper alloy was stir-cast with silicon carbide (SiC) powder to improve various mechanical properties such as increased strength, stiffness, corrosion resistance, wear resistance, and more. Stir-casting involves heating of all composite ingredients to temperatures that promote bonding, then stirring all materials together to diffuse ingredient particles evenly throughout the matrix. Wear testing was conducted utilizing pin-on-disk testing with samples cut from the composite material. Using three input parameters (copper weight percent, silicon carbide

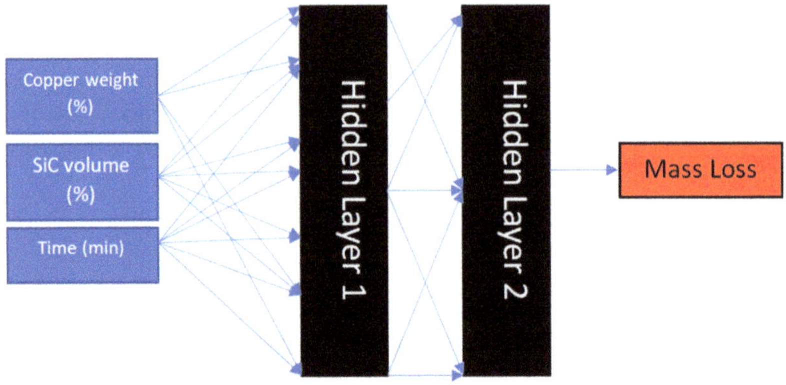

Figure 5.4. Representation of the ANN described by Hayajneh *et al* [30].

volume percent, time), as well as two hidden layers, a back-propagating ANN (shown in figure 5.4) was constructed. The ANN was trained over 100 000 cycles with normalized error starting above 0.0156 and dropping to 0.000 085 by the conclusion of training. The ANN was found to be effective and efficient in predicting mass loss in the wear test conducted on the composite material.

With the increasing focus on environmentally-friendly substances, 'green lubricants', are designed to tackle frictional challenges while maximizing service life and efficiency. Boidi *et al* [31] used radial basis function (RBF), an ML technique, to predict the tribological performance of a certain green lubricant on a variety of surfaces. Values from previous tribological testing, such as slide-roll ratio, geometric surface characteristics, and entrainment velocity were collected. This data was then inputted into the RBF model to train the algorithm to recognize the surface conditions that would lead to different frictional conditions created by the green lubricant. It was discovered that the hardy multi-quadratic radial basis function effectively predicted COF values with an *R*-squared correlation coefficient value of 0.935 using a sample size of 35. However, the researchers noted that the technique was significantly less effective in conditions wherein the surface had shallow and less dense features. Senatore *et al* [32] trained an ANN as well to predict the COF values for piston rings found in common automotive engines. Due to the interaction between piston rings and the cylinder wall being one of the most significant areas for power loss, the researchers felt it necessary to explore options for predicting the friction loss at the interface. Input parameters, such as crank radius, engine speed, oil viscosity, etc were used to train the ANN. Upon completion of training, the ANN was capable of predicting COF values within a correlation coefficient (when compared to calculated values) of 0.9934. ANN output values were found to be slightly higher than calculated values. Figure 5.5 references a portion of the experimentation done, displaying the resulting differences between COF values predicted by the ANN and the traditionally calculated values. It was concluded that the ANN was most successful when tasked with predicting COF values based on pre-determined conditions, allowing for easier problem identification and solving.

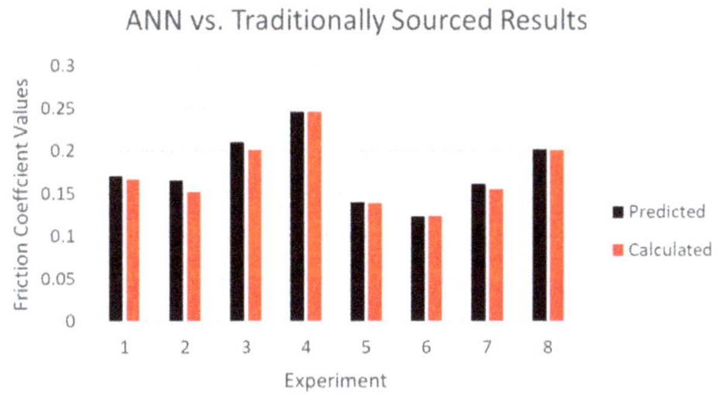

Figure 5.5. Traditionally calculated COF values compared to ANN predictions for top ring/cylinder wall in a SI engineer adapted from [32].

Table 5.1. Accuracy of ANN predictions compared to experimental values adapted from [23].

Material pairing	R^2 result
Test	0.990
Solvent/solvent	0.980
Solvent/polymer	0.963

Similarly, Loh *et al* [33] utilized a feed-forward ANN to predict the dynamic viscosity of oil-based lubricants. A total of 15 different input parameters were used to train the neural network, displayed in table 5.1. The temperature–viscosity relationship, the temperature being an especially important viscosity modifier, was specifically targeted during the training process to improve the network's performance at a variety of heat levels. Two categories of lubricants were analyzed for the study. The first, classified as solvent/solvent, contained such pairing as hydrocarbon–hydrocarbon, hydrocarbon–ester, etc. The second, named solvent/polymer, contained lubricants consisting of diethylene glycol diethyl ether (DGDE) solvent and various polymers (poly butyl methacrylate, etc). These two sets of data, in combination with a third set used to train the ANN before test analysis, were used for experimentation. Table 5.1 displays the R^2 results of the ANN dynamic viscosity predictions. This value represents how effective the ANN was in making the viscosity predictions, compared to found experimental values. It was concluded that even for pairings with very large mole fraction values, the system was highly accurate in its predictive power.

Due to variability in conditions experienced by metals during manufacturing, it can be difficult to accurately predict mechanical and tribological properties without extensive testing. Fragassa *et al* [34] applied ML techniques to predict the hardness of cast spheroidal iron (SGI) and compact graphite iron (CGI) alloys. Three different ANN techniques were utilized, these being random forest (RF), neural network (NN), and the previously discussed *k*-nearest neighbor (KNN) algorithm. Metallographic descriptors such as the quantity of graphite, the quantity of ferrite, the quantity of perlite, the grade of nodularity, and the grade of vermicularity were used as inputs. The predicted values for alloy hardness were outputted from the ANNs. The RF technique is a classifier algorithm taking concepts from binary decision trees and bagging. It was found that the ML techniques were highly suitable for the predictive task, accurately predicting hardness values within a confidence interval of ±15%. It was noted that the process had significant room to be optimized due to several key reasons. Firstly, all code used in the process was openly sourced and easily accessible, with tweaking to better fit the given scenario, it was theorized that accuracy would further increase. Furthermore, different techniques such as the SVM and genetic programming were proposed as potential ways to improve the process.

The design of high wear-resistant polymers is useful due to the continual and growing reliance on polymers worldwide. Short fibers can be easily interfaced with the polymer matrix to offer enhanced mechanical properties. It is well-known that reducing wear decreases the negative effect experienced by the environment from polymer pollution. Zhang *et al* [35] developed a multi-layer feed-forward ANN that was used to predict the specific wear rate and COF of short fiber reinforced polyamide composites. Learning techniques such as Bayesian regularization, Powell–Beale conjugate gradient algorithm, BFGS quasi-Newton method, adaptive learning rate, and Levenberg–Marquardt algorithm were taken into account during ANN construction. Material compositions, mechanical properties, and experiment conditions were input into the neural network, processed through multiple hidden layers, and then output as specific wear rate and the COF. The dataset was increased compared to a previous publication by Zhang *et al* [35], greatly increasing the accuracy of the ANN's predictions. Several conclusions were formed: the size of the dataset greatly affects the predictive accuracy of an ANN, and the ability of an ANN to understand minute relationships between input parameters is the major advantage of ML utilization. Moreover, because internal representations must be adjusted during the training process, using an ANN results in close study and understanding of relationships between input parameters and resulting friction/wear conditions.

Like Zhang, Genel *et al* [36] introduced alumina short fibers to zinc-aluminum alloy, thereby increasing mechanical performance and offering increased resistance to wear. The fibers were interfaced using pressure die-casting, with fibers distributed roughly 200 μm apart in the alloy matrix. Pin-on-disc testing using a rotary tribometer setup was conducted under various loading at a constant speed. Using a classic feed-forward multi-layer ANN, COF and specific wear rate are input from the testing as training data. After training the network to high predictive accuracy, the network was used to predict COF and wear rate. The accuracy of the ANN, when compared to the calculated values, was found to be 94.2% for the wear rate and 99.4% for the COF. It was discovered that increasing the fiber volume fraction decreased the specific wear rate. The researchers concluded that ANN was an excellent analytical tool, saving the team considerable time and money.

Gyurova *et al* [37] used ANN to predict the friction and wear properties of polyphenylene sulfide (PPS). The optimal brain surgeon method (OBS) was applied in order to minimize training time and subsequent cost. OBS works by removing connections within an ANN that are found to be not relevant to the end goal. It was found that values such as the material composition, test conditions, tensile modulus, and strain were correlated with the most change in wear performance. Using these values as input to the network improved the predictive quality. The ANN was successful in carrying out accurate predictions on the COF and specific wear rate of the polyphenylene sulfide composite.

5.6 Conclusions

This review summarized key works involving ML in tribology, as well as giving a basic introduction to what ML is. The compilation of ML applications in the field of tribology displays the power and efficiency available through techniques. The methods discussed enabled users to spend significantly smaller amounts of time and money on experimentation, leaving more time for analyses. Evidently, ML is most suitable for predictive tasks, simply based on the volume of research involving this task. However, it is also valuable in monitoring large mechanical systems, wherein it may be difficult to monitor internal wear. In these cases, ML has been proven to be effective in discovering problems and enabling system diagnostics, before catastrophic failure occurs. Furthermore, material design utilizing ML enables researchers to quickly develop materials with specific desired properties. The field of ML will only increase in size, efficiency, and capability, giving researchers a valuable tool to utilize.

References

[1] McCulloch W S and Pitts W 1943 A logical calculus of the ideas immanent in nervous activity *Bull. Math. Biophys.* **5** 115–33

[2] Rosenblatt F 1958 The perceptron: a probabilistic model for information storage and organization in the brain *Psychol. Rev.* **65** 386

[3] Krogh A 2008 What are artificial neural networks? *Nat. Biotechnol.* **26** 195–7

[4] Rumelhart D E, Hinton G E and Williams R J 1986 Learning representations by back-propagating errors *Nature* **323** 533–6

[5] Rumelhart D E, Widrow B and Lehr M A 1994 The basic ideas in neural networks *Commun. ACM* **37** 87–93

[6] Schapire R E 1990 The strength of weak learnability *Mach. Learn.* **5** 197–227

[7] Schapire R E 2003 The boosting approach to machine learning: an overview *Nonlinear Estimation and Classification* (New York: Springer) pp 149–71

[8] Menezes P L, Nosonovsky M, Ingole S P, Kailas S V and Lovell M R 2013 *Tribology for Scientists and Engineers* (Berlin: Springer)

[9] Mohri M, Rostamizadeh A and Talwalkar A 2018 *Foundations of Machine Learning* (Cambridge, MA: MIT Press)

[10] Boser B E, Guyon I M and Vapnik V N 1992 A training algorithm for optimal margin classifiers *Proc. Fifth Annual Workshop on Computational Learning Theory* 144–52

[11] Noble W S 2006 What is a support vector machine? *Nat. Biotechnol.* **24** 1565–7

[12] Whitley D 1994 A genetic algorithm tutorial *Stat. Comput.* **4** 65–85

[13] Wan Z, Wang Q-D, Liu D and Liang J 2021 Discovery of ester lubricants with low coefficient of friction on material surface via machine learning *Chem. Phys. Lett.* **773** 138589
[14] Bhaumik S, Pathak S, Dey S and Datta S 2019 Artificial intelligence based design of multiple friction modifiers dispersed castor oil and evaluating its tribological properties *Tribol. Int.* **140** 105813
[15] Edla S, Krishna A, Karthik G, Arif M M and Rani S 2021 Potential use of transesterified vegetable oil blends as base stocks for metalworking fluids and cutting forces prediction using machine learning tool *Biomass Convers. Biorefinery* 1–12
[16] Bharadwaj P V, Jeevan T, Suvin P and Jayaram S 2019 Prediction of surface roughness and coefficient of friction using artificial neural network in tribotesting of bio-lubricants *Appl. Mech. Mater.* **895** 52–7
[17] Bhaumik S, Mathew B R and Datta S 2019 Computational intelligence-based design of lubricant with vegetable oil blend and various nano friction modifiers *Fuel* **241** 733–43
[18] Humelnicu C, Ciortan S and Amortila V 2019 Artificial neural network-based analysis of the tribological behavior of vegetable oil–diesel fuel mixtures *Lubricants* **7** 32
[19] Mujtaba M, Masjuki H, Kalam M, Ong H C, Gul M, Farooq M, Soudagar M E M, Ahmed W, Harith M and Yusoff M 2020 Ultrasound-assisted process optimization and tribological characteristics of biodiesel from palm–sesame oil via response surface methodology and extreme learning machine-Cuckoo search *Renew. Energy* **158** 202–14
[20] Dey S, Sultana N, Kaiser M S, Dey P and Datta S 2016 Computational intelligence based design of age-hardenable aluminium alloys for different temperature regimes *Mater. Des.* **92** 522–34
[21] Vinoth A and Datta S 2020 Design of the ultrahigh molecular weight polyethylene composites with multiple nanoparticles: an artificial intelligence approach *J. Compos. Mater.* **54** 179–92
[22] Umeda A, Sugimura J and Yamamoto Y 1998 Characterization of wear particles and their relations with sliding conditions *Wear* **216** 220–8
[23] König F, Sous C, Chaib A O and Jacobs G 2021 Machine learning based anomaly detection and classification of acoustic emission events for wear monitoring in sliding bearing systems *Tribol. Int.* **155** 106811
[24] Mokhtari N, Pelham J G, Nowoisky S, Bote-Garcia J-L and Gühmann C 2020 Friction and wear monitoring methods for journal bearings of geared turbofans based on acoustic emission signals and machine learning *Lubricants* **8** 29
[25] Kankar P K, Sharma S C and Harsha S P 2011 Fault diagnosis of ball bearings using machine learning methods *Expert Syst. Appl.* **38** 1876–86
[26] Altay O, Gurgenc T, Ulas M and Özel C 2020 Prediction of wear loss quantities of ferro-alloy coating using different machine learning algorithms *Friction* **8** 107–14
[27] Xu X, Zhao Z, Xu X, Yang J, Chang L, Yan X and Wang G 2020 Machine learning-based wear fault diagnosis for marine diesel engine by fusing multiple data-driven models *Knowl.-Based Syst.* **190** 105324
[28] Borjali A, Monson K and Raeymaekers B 2019 Predicting the polyethylene wear rate in pin-on-disc experiments in the context of prosthetic hip implants: deriving a data-driven model using machine learning methods *Tribol. Int.* **133** 101–10
[29] Thankachan T, Soorya Prakash K and Kamarthin M 2018 Optimizing the tribological behavior of hybrid copper surface composites using statistical and machine learning techniques *J. Tribol.* **140** 031610

[30] Hayajneh M, Hassan A M, Alrashdan A and Mayyas A T 2009 Prediction of tribological behavior of aluminum–copper based composite using artificial neural network *J. Alloys Compd.* **470** 584–8
[31] Boidi G, da Silva M R, Profito F J and Machado I F 2020 Using machine learning Radial Basis Function (RBF) method for predicting lubricated friction on textured and porous surfaces *Surf. Topogr.: Metrol. Prop.* **8** 044002
[32] Senatore A and Ciortan S 2011 An application of artificial neural networks to piston ring friction losses prediction *Mech. Test. Diagn.* **1** 7–14
[33] Loh G, Lee H-C, Tee X, Chow P and Zheng J 2020 Viscosity prediction of lubricants by a general feed-forward neural network *J. Chem. Inf. Model.* **60** 1224–34
[34] Fragassa C, Babic M and dos Santos E D 2020 Machine learning approaches to predict the hardness of cast iron *Tribol. Ind.* **42** 1–9
[35] Zhang Z, Friedrich K and Velten K 2002 Prediction on tribological properties of short fibre composites using artificial neural networks *Wear* **252** 668–75
[36] Genel K, Kurnaz S C and Durman M 2003 Modeling of tribological properties of alumina fiber reinforced zinc–aluminum composites using artificial neural network *Mater. Sci. Eng.* A **363** 203–10
[37] Gyurova L A, Miniño-Justel P and Schlarb A K 2010 Modeling the sliding wear and friction properties of polyphenylene sulfide composites using artificial neural networks *Wear* **268** 708–14

IOP Publishing

Multifunctional Bio-Based Lubricants
Synthesis, properties and applications
Ashish K Kasar, Arpith Siddaiah and Pradeep L Menezes

Chapter 6

Environmentally friendly grease

Ashish K Kasar, Reuel Baluyut, Jonfred Christian Diamzon, Morgan Lester, Rich Milani, Guillermo Ramirez and Pradeep L Menezes

Grease is a solid or semi-solid lubricating agent formed by the dispersion of a thickening agent in a liquid lubricant. After usage, impropriate disposal of toxic grease materials can be extremely harmful to soil or water. The disposal of grease is paramount, with extant methods being re-refining, landfill disposal (requiring water removal from the grease prior to disposal), and incinerating at extremely high temperatures. Therefore, environmentally friendly greases are being developed. This chapter discusses three types of discernable 'environmentally friendly' greases: vegetable oils, polyalkylene glycol (PAG), and synthetic esters. Among these three, excluding vegetable oils, which were found to be uncompetitive with PAG and synthetic esters, both were found to have 'low' aquatic toxicity and bioaccumulation. PAG can have the best biodegradability to break down in the environment, partial to its ability to be constructed with a definitive timeline of breakdown with exposure to moisture. Vegetable oils are naturally the grease/oil with the highest natural breakdown rate, with synthetic esters being the least of the three, in terms of ability to degrade in the environment. In addition to the properties of environmentally friendly grease, a market analysis is provided in the chapter that suggests PAG holds a majority in the market since it is more economical and useful in high-temperature applications.

6.1 Introduction

When two or more objects rub against one another, a little bit of material gets shaved off. This, along with the generation of heat due to friction, contributes to the wear of the components, which would eventually lead to a mechanical failure of the system. History suggests that lubricant paves the way for inventions that create modernization. Among different lubricants, grease is an incredibly important substance used in a vast array of different mechanical environments. Whenever a thick lubricant is required, grease is the favorable lubricating material. Lubricating grease is defined as a 'semifluid product of dispersion of a thickening agent in liquid lubricant'. The components of grease consist of oil, thickeners, and additives. The majority of the grease is made up of a base oil which can be mineral, synthetic and vegetable oil or a combination of these oils. The thickening agents can be simple or complex metal soaps. If they are not metal based soaps, then they can be non-soap thickeners. The additives in lubricating grease are typically used to enhance the properties, such as inhibiting corrosion, etc.

The purposes of grease are not only to lubricate, but to prevent corrosion and even dampen movements. When a mechanical system needs to be lubricated for a long time, grease is used because it stays within its constant properties, it does not develop a film of something similar. This allows grease to be used in parts that cannot be accessed frequently for the chance to re-grease. Grease can also be used in high temperature or high-pressure situations because it maintains its thicker property [1]. Grease is a semi- solid substance that becomes liquid when it starts moving and has a higher viscosity than other liquid lubricants. Grease is used to lubricate mechanical systems that cannot be automatically supplied with oil, such as tools, ball and roller bearings, shafts, chains, and other rotating parts. Mineral oil-based greases are widely used that are composed almost entirely of refined mineral oil, processed in a way that increases viscosity to an extreme level. Unfortunately, mineral oil is not only quite toxic for humans, being listed as a group 1 carcinogen, but its properties and lack of breakdown do not allow for easy deconstruction by environmental means. Grease is inherently designed to be waterproof and to resist elemental and thermal breakdown, which makes the task of creating a lubricant that is 'environmentally friendly' quite difficult to achieve. In the next section, major types of greases are discussed with respect to their applications.

6.2 Types of grease and their application

The primary purpose of grease is to reduce friction within mechanical systems. These systems can include gears, bearing, railings, etc. Grease is a semi-solid lubricant that is essential and vital for machinery to run as smoothly and efficiently as possible. The application and types of grease are vast depending on the machinery and even the type of industry the grease is applied to.

Within the automotive industry, multipurpose grease works well for chassis components. This is primarily because heat does not become an issue. Although it is called a multipurpose grease, this grease is primarily made up of calcium sulfonate [2]. In the food industry, equipment and machinery for food processing plants require specialized lubricants. The grease used in these machines needs to be food grade. Not only should the greases have requirements with oxidation stability, load capacity, and corrosion protection, but they also need to meet health requirements because the items going through the machines are to be consumed and digested by consumers [3]. This is why aluminum complex greases are great for food handling machinery. Aluminum grease is a great choice outside of the food industry because it can resist water and has better oxidation stability, as well as heat resistance—meaning it can operate at higher temperatures. Due to these great properties, the price of aluminum grease does increase [4, 5].

Bentone is typically used in low-speed bearings because of its high viscosity. One use for bentone grease is to improve load-carrying capacity and electrical conductivity in certain machinery [6]. In the industry, bentone also goes by 'no-melt' grease for its good water resistance, and high temperature applications. Due to its claylike structure, bentone grease does not work well with other greases [7]. Other characteristics that make bentone a great lubricant choice is its ability to adhere to surfaces, wear reduction, and shear stability [6]. The other widely used greases are lithium and lithium complex grease, barium complex grease, polyurea grease, and sodium grease. Understanding the types of grease and their applications is vital to be able to take a step back and approach grease in a new light.

6.3 Environmentally friendly grease

For lubricants to be taking a stance to become part of a green initiative, one major characteristic is being extracted from a renewable source, meaning that the source should be produced from harvested products in a relatively short time. These should be products such as animals and vegetables. However, these types of sources do not produce stable lubricants, which means that their instability makes them biodegradable, which is in turn an environmental advantage.

The creation of an environmentally friendly, biodegradable grease has been a sought-after research breakthrough for many years. However, the main concern for such a grease is its effectiveness and its cost to manufacture. In 2009, a biodegradable grease was developed by researchers from the University of Huelva [8]. These types of grease are called 'oleogels' and are made with natural ingredients designed to degrade without damaging the environment over time. It is important to notice about this new grease is that the cellulose derivatives need simpler technology to manufacture, as well as being easier to process. However, a common problem with these kinds of grease is that they have lower mechanical stability and higher leakage tendency.

One 'green' grease uses cellulose derivatives from plants and Ricin oil as a lubricant base. Another uses Caster oil and cellulose derivatives. In 2015, researchers from the Bio Fuels Division at the CSIR-India developed a biodegradable multipurpose grease made from Jatropha vegetable residual oil, a lithium soap and an undisclosed additive [9]. Condat Corp has developed a multitude of biodegradable greases for specific fields of applications; however, what they are actually made of is undisclosed [10].

Although some lubricants can be reused, most of the other options are non-reusable due to degradation and contamination. Early environmentally acceptable lubricants were made from bio-based materials or were biodegradable, mostly formulated using vegetable oil-based fluids. The effects of this early environmentally friendly lubricant can be visually seen due to its jelly-like structure at low temperatures. Along with its extensive formulation, the cost of this early-stage product did not come with a good price. Fortunately, many options are available today through genetically improved vegetable oils or high-performance synthetic fluids so that higher-performing products can be formulated to overcome the low- and high-temperature properties of the early products [11]. With this new technology, researchers and manufacturers found a new and green way of mass producing environmentally friendly grease with less ecotoxicity and bioaccumulation.

Another way to be environmentally friendly is to reduce the amount of natural resources used for production [11]. The oil industry uses expensive machinery to dig fossil fuel from the ground. By reducing the old style of getting crude oil, we can reduce air pollution and oil fracking. Oil fracking can cause shifts in geological surfaces, it also causes settling and small earthquakes in nearby areas [12]. As for the atmosphere, fracking causes the release of trapped gasses underground, such as methane.

In reality, a successful environmentally friendly grease program is the responsibility of the end user. It is they who decides how to dispose of the residual waste and containers the grease and lubrication came in. A world-class lubrication reliability program improves equipment reliability and uptime with the fringe benefit of creating a holistic green lubricant program [11]. As for companies, switching to green initiatives would greatly reduce the cost of production and will increase profitability. During this current generation, a green initiative would produce a significant number of consumers who believe in a healthy and safe environment for the future generations to come, thus, making the market for environmentally friendly grease a better market economically and environmentally. There are a number of different aspects of a lubricant that classify it as biodegradable. These aspects include biodegradability.

6.3.1 Biodegradability

Biodegradability is a measure of the breakdown of a chemical (or a chemical mixture) by microorganisms. There are two types of biodegradation, primary and ultimate. Primary biodegradation is the loss of one or more active groups in a chemical compound that renders the compound inactive with regards to a particular function. Primary biodegradation can turn a toxic compound into a less toxic or even non-toxic compound. The other form of biodegradation is ultimate biodegradation, which can also be referred to as mineralization, and is the process of a chemical compound being converted into a carbon dioxide, a water, or a mineral salt.

Biodegradation is also defined by two operational properties. These two operational properties include inherent biodegradability and ready biodegradability. A compound that is considered inherent is one that displays evidence of biodegradation in a biodegradability test. A compound is readily biodegradable if one fraction of the compound is biodegradable.

There are several test methods to determine if a compound is biodegradable. These include the Organization for Economic Cooperation and Development (OECD), OECD 301B [13], American Society for Testing and Materials (ASTM), ASTM D-5864 [14], and Coordinating European Council (CEC), CEC L-33-A-93 [15]. Parameters of the these tests are listed in table 6.1. Both OECD 301B and ASTM D-5864 perform a test to determine if 60% of the compound can be converted into carbon dioxide. The CEC method determines if 80% is biodegradable. The CEC method is considered to be a less stringent test as it does not define a compound as primary or ultimate biodegradability.

Pursuing an environmentally friendly alternative has been a desirable goal for many industries. Most eco-friendly greases use vegetable oil as their base with a

Table 6.1. Different biodegradability tests and their parameters.

Test type	Test name	Measured parameter	Pass level	Method
Ready biodegradability (a substance is considered to be inherently biodegradable using any of these tests if it shows >20% biodegradability within the test duration)	DDAT	DOC	>70%	OECD 301A
	Strum test	CO_2	>60%	OECD 301B
	MITI test	DOC	>70%	OECD 301C
	Closed bottle test	BOD/COD	>60%	OECD 301D
	MOST	DOC	>70%	OECD 301E
	Sapromat	BOD/COD	>60%	OECD 301F
	Strum test	CO_2	>60%	ASTM D-5864
	Shake flask test	CO_2	>60%	EPA 560/6–82–003
	BODIS test	BOD/COD	>60%	ISO 10708
Hydrocarbon degradability	CEC test	Infrared spectrum	>80%	CEC L-33-A-934
Screening tests (semi-official)	CO_2 headspace test	CO_2	>60%	ISO 14593

thickening agent to make it into a grease instead of lubricant oil, among other ingredients. For biodegradable grease, the oils discussed in this topic are palm oil, rapeseed oil, and castor oil. Most conventional greases use mineral oil thickened with a metallic soap. Lithium-based grease will be the focus for conventional grease, which is the kind of grease widely used in the world. Needless to say, eco-friendly grease is environmentally friendly, while conventional grease is not. Mineral oils are dangerous to the environment and will pollute both water and soil. Bio-based greases are about sixty to eighty percent biodegradable. In addition to the fact that petroleum and crude oil are not inexhaustible resources, waste products from these resources present a hazard to human health and the environment.

6.3.2 Aquatic toxicity

In addition to the biodegradability aspect of an environmentally friendly lubricant, aquatic toxicity also plays a part in determining if grease is safe. Similar to the biodegradability tests, aquatic toxicity can be determined using a series of different tests. These tests are listed in table 6.2. The most common aquatic toxicity tests for assessing environmentally friendly lubricants are the 72-h growth test for algae (OECD 201), the 48-h acute toxicity test for daphnia (OECD 202), and the 96-h toxicity test for fish (OECD 203).

The petroleum-based oils have higher toxicity to biota compared to the other compounds. This is a result of the more rapid breakdown of petroleum-based oils once they are in the water. The toxicity levels of petroleum-based oils depend on additives used in formulations and metabolites produced in biodegradation. The use of additives depends on the base oil as well as the intended function of the lubricant. The toxicity level of the base oils for greases are mentioned in table 6.3. Usually, the more toxic compounds are also compounds with poor degradability. The overall

Table 6.2. Various aquatic toxicity tests.

Test title, with species	Test Number
Growth inhibition test, alga	OECD 201
Acute immobilization test, *Daphnia* sp.	OECD 202
Acute toxicity test, fish	OECD 203
Prolonged toxicity test: 14-day study, fish	OECD 204
Respiration inhibition test, bacteria	OECD 209
Early-Stage toxicity test, fish	OECD 210
Reproduction test, *Daphnia magna*	OECD 211
Short-term toxicity test on embryo and sac-fry stages, fish	OECD 212

Table 6.3. Toxicity levels of different base oils for grease production with their source.

Lubricant base oil	Base oil source	Toxicity
Mineral oil	Petroleum	High
polyalkylene glycol (PAG)	Petroleum—synthesized hydrocarbon	Low
Synthetic ester	Synthesized from biological sources	Low
Vegetable oils	Naturally occurring vegetable oils	Low

product toxicity can be lowered by switching to a biologically sourced base oil as well as using low toxicity additives.

6.3.3 Bioaccumulation

The propensity of a substance to bioaccumulate is another property of a lubricant that is often considered in the qualification of a product as an environmentally friendly lubricant. Bioaccumulation is the accumulation of chemicals within the tissues of an organism over time. The longer that the organism is alive, the greater the accumulation of chemicals that occurs. If the chemical has a slow degradation rate or low depuration rate within an organism, concentrations of that chemical may build up in the organism's tissues. This may eventually lead to adverse biological effects. This is why it is better to use compounds that do not bioaccumulate. Even though total removal of bioaccumulation may not be possible, it is possible to use chemicals that have a slower bioaccumulation rate.

The bioaccumulation trait of a compound is directly proportional to its water solubility. These chemicals that are responsible for water solubility include most of the compounds found in the creation of base oil in environmentally friendly lubricants. The water solubility of a compound depends on the type of atoms in the molecule. Carbon and hydrogen have the lowest solubility in water, and these are often found in alkanes which make up almost all of the current base oil in lubricant formulas. When oxygen atoms are present in a molecule, it will generally increase the water solubility, which will then decrease bioaccumulation. When a compound includes oxygen, this also tends to cause that compound to degrade at a faster rate. There are a lot of different base oils that are formulated using carboxylic acids because they increase solubility and degradation, which limits their ability to bioaccumulate in comparison to alkane-based oils.

A topic that has been discussed for a long time is the size of the molecule and whether it is bioaccumulated because it is too big to pass through the membrane of a cell. As a result, the size of the molecule is taken into consideration when designing and manufacturing an environmentally friendly lubricant. There have been a lot of discussions and proposed ideas about the point at which a chemical is able to be bioaccumulated. There is a lot of data that evaluate this kind of situation and there was a substantial connection between the log K_{ow} and log BCF. log K_{ow} is the logarithm of the partitioning coefficient of a substance in *n*-octanol and water, and the log BCF is a measure of the bioconcentration from water into aquatic organisms.

Some environmentally friendly lubricant labeling programs like the European Eco-label demand proof that a compound is not bioaccumulative. There are many ways to provide this information, including measuring the log K_{ow} or the log BCF. Two very common tests for bioaccumulation are the OECD 117 and OCED 107. In these tests, the compound in question is combined with octanol and water. Then, using gas chromatography or an infrared laser, the dissolution is observed. The bioaccumulation of the compound is measured by finding its partition coefficient. Compounds that bioaccumulate usually dissolve in the octanol instead of dissolving in water. Compounds that possess a higher log K_{ow} value tend to have a higher chance of bioaccumulation [16]. Partition coefficients in aquatic compounds are typically measured on a log scale from zero to six. Compounds with a log K_{ow} value that is lower than three tend to not bioaccumulate, while compounds with a log K_{ow} value of greater than three tend to bioaccumulate. Saltwater has an effect when it comes to testing for bioaccumulation in the marine setting. This effect might cause an increase in bioaccumulation potential in certain compounds.

6.3.4 Labeling

In order to increase public knowledge and decrease the aversion for lubricants, national and international labeling programs were developed. These labeling programs would determine if lubricants are environmentally friendly.

The first national labeling program that was developed was Blue Angel. This program was developed in 1988 in Germany [17]. This program classifies lubricants as hydraulic fluids, lubricant oils, and greases. For a substance to qualify for certification, it must exemplify different characteristics. These characteristics include biodegradability, low toxicity to aquatic organisms; non-bioaccumulative; and no dangerous components. In addition to these characteristics, substances must also pass technical requirements in order to receive the Blue Angel certificate. These characteristics depend on the class of the lubricant, and unlike other labeling programs, Blue Angel does not deal with renewability. Generally, Blue Angel certification is pretty strict, and many lubricants do not receive this certification. The majority of substances that do receive this certification are hydraulic fluids.

Like Blue Angel, Swedish Standard is another national labeling program that mainly deals with hydraulic fluids and greases. This program tests for aquatic toxicity, biodegradability, and sensitizing properties. The Swedish Standard tests for biodegradability by using ISO testing procedures which has different requirements depending on the class of the substance. Unlike other programs, Swedish Standard was developed with both the government and industry in mind. Swedish Standard has more listed lubricants, more specifically hydraulic fluids, than any other national labeling program.

Nordic Swan was the first international labeling program for environmentally friendly lubricants [18]. The program comes from a region that includes Norway, Denmark, Sweden, Finland, and Iceland. Nordic Swan deals with hydraulic oils, two-stroke oil, grease, transmission fluid, and gear oil. To qualify for certification from the Nordic Swan labeling program, a lubricant must pass tests in biodegradability, aquatic toxicity, technical performance, and renewability. The requirements for renewability are some of the stricter requirements of all the labeling programs. For example, hydraulic fluid, transmission fluid, gear oil, and grease must be at least 65% renewable. Two-stroke oil must be 50% renewable. As a result of these high standards, few lubricants receive the Nordic Swan label [18].

The European Eco-Label is considered by many to be one of the more popular and accepted labeling programs. This program was established in 2005 and dealt with hydraulic fluids, greases, and two-stroke oils [19]. To receive certification from European Eco-Label, a lubricant must pass tests for biodegradability, aquatic toxicity, bioaccumulation, and toxic substances. A lubricant cannot have any trace of dangerous materials that could lead to potential human or environmental harm. Some qualities include explosiveness, flammability, carcinogenic potential, volatility, potential to cause birth defects, and others. In terms of bioaccumulation and biodegradability, hydraulic fluid must be at least 90% biodegradable, while greases and two-stroke oils must be 75% biodegradable. For a lubricant to be considered as bioaccumulative, it must meet one of the following conditions: it must have a molar mass of 800 g mol^{-1} or greater, it must have a K_{ow} value that is lower than three or greater than seven, or it must have a BCF of 100 L kg^{-1} or lower. Lubricants also cannot contain halogenated organic compounds, nitrite compounds, metals or metallic compounds. Finally, in order for a lubricant to receive a certificate from European Eco-Label, 50% of hydraulic and two-stroke oils must be composed of renewable materials, while 45% of greases must contain renewable materials. In this case, renewable materials are things like animal fat or vegetable oils [20].

The final labeling program is called OSPAR (Oil Spill Prevention, Administration and Response), which deals with chemicals that are used and discharged from offshore oil facilities in the North Sea [21]. These chemicals, including well chemicals, are released during regular use. The Convention for the Protection of the Marine Environment of the North-East Atlantic is leading the way for protecting the marine environment in the North Atlantic. Operations of this convention are overseen by OSPAR, which is made up of 15 different governments. OSPAR regulates chemicals that are released into the environment in regard to biodegradation, bioaccumulation, and toxicity. These qualities are some of the more common regulations, as well as being the most appropriate, that are used by different programs around the world. OSPAR evaluates environmental performance for persistence, which is biodegradation over a 28-day period, bioaccumulation, and aquatic toxicity for four native species, algae, copepods, sediment re-workers, and bottom-dwelling fish. Testing must be done by an approved third-party organization.

6.4 Grease and regulation

The exponential speed of human progression to revolutionize development in various sectors came with a great price. As a result of the advancement in automotive, industrial, marine and aviation, 'petroleum has been depleting worldwide' [22]. The market for grease and lubricant has been flat and may continue to be flat for the foreseeable future. This is due to the fact that suppliers and makers have no contingency plan for making reusable and biodegradable grease and lubrication. However, the demand for synthetic and biodegradable grease has grown in recent years, making the market stable and the industry more sustainable and suitable for a new breed of consumers.

Nowadays, 'an increase in interest in the assessment of the impact of lubricating oils on the environment and health motivates research centers and industry to create new technologies for the production of fully biodegradable, lubricating oils of natural origin' [23]. The new generation of young adults has keenly embraced the green initiative. The new generation takes an active role in saving the planet by reducing waste in any form. Although the majority of the green initiative right now can only be seen in cleaning the streets, oceans and lakes, and banners promoting cleanliness and less toxic waste to avoid polluting oceans and lakes that inhabit thousands of species, in the near future, they will be the driving force of the world. Part of the hardship in tackling the situation and implementing environmentally-friendly grease into the market is that the vast majority of companies that make grease refuse to switch to a greener and more biodegradable solution. This is because it will cost them money, and they will likely lose money in the first year or two after switching just by investing in new machinery and implementing new processes and chemicals.

The current regulation in petroleum-based grease and lubrication is also a part of the problem. 'Governments, in cooperation with industry, develop standards, regulations, and procedures to reduce the risk of accidents and leaks. However, in the case of streams introduced into the environment in the form of oil mist, such limitations are physically impossible, hence the need to introduce legal regulations regarding the composition of the indicated oils' [24]. When lubricating oil is released into the environment, it poses a major threat to the ecosystem. The biodegradability of mineral oils is very low. In the natural environment, oil of petroleum origin creates primary hazards for sawing operators, but also secondary hazards due to the accumulation of oils in plant, animal, and groundwater tissues. Lubricating oils produced from crude oil are also a very significant threat to aquatic ecosystems. This is due to the process in which lubricating oil is disposed of. Grease and lubrication are emitted in the form of mist and micro drops that blend into the environment. This is why transitioning to eco-friendly and biodegradable grease and lubricant should be imposed by the government to minimize the damage large factories cause to the environment.

In the European Union, the council that oversees this problem developed a test that establishes a method for biodegradation that accounts for testing grease emission to the environment. 'OECD 301 A–F methods assess biodegradability by subjecting the test oil to specific microorganisms, most often for 28 days. Tests allowing a degradation value of 60% at a given time (28 days) may be inadequate for lubricating oils, because the remaining 40% of substances can be subjected to natural biodegradation over even several hundred years' [23]. Although the test is reduced in scale, it does not completely solve the emission problem.

Since petroleum-based oil and lubricants greatly impact everyday use, it is worth investing in a greener and environmentally friendly way in which grease and lubricants can be disposed of. 'Oils produced from crude oil have a negative impact on the environment, cause serious contamination of soils, groundwater, and can accumulate in plant tissues as well as terrestrial and aquatic animals' [23]. Not only does it impose a significant health hazard on the environment and to the people, it is also the cause of many allergic reactions due to the changes it causes to the environment we live in. For a biodegradable solution to work, the people in charge should provide a complete test that will totally confirm the environmentally friendly grease's biodegradability. New laws and regulations should be given to the companies that make grease and lubricants to provide proper guidelines and rules on the design verification of the product. Legal regulations should also be specific with accuracy in explaining the outcome needed for the grease and lubricant to be considered biodegradable.

6.4.1 Disposal of mineral and synthetic based lubricants

It is estimated that approximately 37 million tons of lubricants are used in the world each year. Approximately 10% of these lubricant products are used in systems where the waste lubricant is discarded directly into the environment. It is also estimated that 13%–32% of used lubricants are returned to the environment in a state where they are essentially unchanged [25]. When these used lubricants do not get discarded properly, the effects are seen in multiple areas of life. These areas can include environmental pollution, which then can have health impacts on humans as well as economic impacts.

Ensuring safe, efficient and economical disposal starts with the collection and the handling of used grease. The producer of the waste must make sure that they store said waste in a container that is suited to do so. Another responsibility of the producer is to make sure the collector is licensed to collect the waste. Vehicles that are used to transport waste must be suited to do so in order to avoid spills. According to the Waste Act 2008 (Act No. 59 of 2008) and the Act on Transport of Dangerous Goods, hazardous materials must be transported as outlined in the act [25]. The container that is most commonly used to transport the waste is the 200 L drum. These containers must be covered, kept outdoors, and stored away from heaters.

When it comes time to dispose of the used grease, there are many guidelines that must be followed. Unfortunately, oftentimes grease ends up in our environment. Sometimes people dispose of grease incorrectly without even realizing it. This can happen when they pour excess grease or lubricant on the ground or in a spillway that leads to sewage. They think that pouring it on grass or in bushes is harmless, but it can often find its way into sewage plants from rainwater lifting and carrying the substances. Disposing grease at a landfill site is a popular option, but the water must be drained from the grease. When water is not separated from the substance, it causes accidental runoff and leakage.

When the grease is dumped in landfill, there is a process that occurs to make sure it is treated properly. One method is called reprocessing, which produces a fuel oil with low basic sediment and water content. This prevents the fuel from clogging burners and will not cause sediment build up. This process filters and removes coarse solids from the substance. This can be achieved by processes like settling, centrifugation or filtration.

Another way to handle used grease is the process of re-refining. Re-refining distills the used lubricant to make a high-quality base stock for more lubricants. The use of this method has increased tremendously in developed countries, reaching up to 50% of some countries' need for lubricating oil [25]. Re-refining uses solvent extraction, which is then followed by clay or acid treatment. Vacuum distillation produces less pollution and is more cost-effective. The result of this process is compacted in thick plastic sheets and then disposed of into landfills.

The final process that is used to dispose of used grease is complete destruction. This is used when the grease is just too contaminated. Hazardous waste incinerators take care of the waste rather easily. However, when these are not an option, the waste is taken to a cement factory and burned at about 2000 °C. This is hot enough to destroy organics as well as neutralizing acid compounds. These factories have to constantly monitor the gas emissions of the furnaces to make sure they stay within air quality standards.

6.5 A comparison between environmentally friendly grease and conventional grease

The development of a renewable source of semi-solid lubricant, called grease, has been a major hot topic in recent years. As much as humanity has advanced, it is simply not sustainable to rely on a finite ingredient for grease, which is crude petroleum oil. Not only that, our consumption of these finite sources for lubrication is causing damage to the planet we humans reside in. The most commonly known effect of damaging the Earth due to our constant reliance on mineral oil is climate change, which, for most people, is a good enough reason to develop a more sustainable, renewable, or 'green' kind of grease.

The comparison between environmentally friendly grease and conventional grease is made in this section with the focus being their performance. Conventional greases are used in nearly every application where there are moving parts and lubrication is needed. They are, therefore, considered an all-purpose grease. Environmentally friendly greases are competing for the title of 'all-purpose,' and so, at the very least, they must perform as well as, if not better than, conventional greases, that is, lithium-based grease.

The main ingredient of a lubricating grease that causes it to be 'environmentally unfriendly' is mineral oil, a highly refined petroleum distillate. Biodegradability is defined as the ability of a substance to completely revert to a basic set of substances, mainly methane, water, mineral salts, and carbon dioxide. The biodegradability of mineral oil is specifically demonstrated to be between 15% and 35% [26], where a true biodegradable substance should show at least 50% biodegradability. In order to have an 'environmentally friendly' grease, then, the biodegradability must be higher than 50%. The substance must also produce excellent lubrication properties and resist thermal breakdown for a substantial or planned time frame.

The other important consideration is that the lubricant being designed will make an impact on the lubricant industry as a whole. The meaning behind this sentiment is this: if a lubricant affects the environment positively or neutrally but is hardly used, then it will not have a demonstrable effect on the damage done by more toxic greases/lubricants. One of the major areas of environmental impact does not occur from what someone would assume to be a main factor, such as cars and trucks, but from acceptable fluid loss systems such as marine transport ships, which have a large, allowable amount of lubricant to be expelled into the environment. Now that the definitions and requirements for an 'environmentally friendly' grease/lubricant have been stated, it's time to start delving into some of the alternatives to toxic petroleum distillate lubricant usage (or perhaps a petroleum distillate that breaks down more effectively or desirably). The three most common biodegradable oils are vegetable oils, synthetic esters, and PAG.

Vegetable oils can widely be considered the most 'environmentally friendly' of the three biodegradable oil types; they are fully organic in nature, break down with heat and oxidation easily, and are generally neutral or even beneficial to flora and fauna. Unfortunately, because of their organic nature, vegetable oils break down too easily under heat and friction and are not stable enough to be used in any heat engine usage.

Synthetic esters are an excellent grease type that has become ubiquitous in use and production. This grease type is not only environmentally friendly but has high heat tolerances and excellent breakdown rates. Uses include aircraft oil, gear lube, and heavy-duty hydraulic applications. Synthetic esters are relatively expensive, however, compared to the cheap and readily available vegetable oil.

The final environmentally friendly grease type is PAG, which is a synthetic lubricant derived from the polymerization of ethylene or propylene oxide [27]. Also, it is suprior to vegetable oils in terms of lubricity, heat tolerance, and general breakdown. This grease type can be specifically designed to be water or oil soluble, which makes it an excellent base lubricant for any biodegradable application.

6.5.1 Current environmentally friendly grease application and performance

Considering the minor uses of vegetable oils for industrial applications, this section will focus on synthetic esters and PAG. To begin, the industrial and machine applications for synthetic esters are substantial, given by their great lubricity, great biodegradability, and excellent oxidative stability, and though lacking in hydrolytic stability (the resistance to bond to water molecules, similar to oxidation), that particular aspect can be engineered out with the help of stabilizing additives. Synthetic esters also benefit from their base materials, which are voluminous and allow for high volume and low-cost production, consisting of carboxylic acid and alcohol [28]. The process that creates synthetic esters is also reversible, aiding in the biodegradability of the lubricant. This reversible process is aided by the presence of water, a double-edged sword in terms of longevity of the grease, and the helpful biodegradability. Additionally, these synthetic esters are incredibly thermally stable. The flash, smoke, and fire points of synthetic esters are incredibly high [28]. The addition of industry-standard stability additives greatly increases the useability and longevity of synthetic esters, high above their previous ability making them an excellent substitute for conventional lubrication methods.

Next, a comparison is made between synthetic esters and the other frontrunner for environmentally friendly grease-PAG. To begin, there are two distinct types of PAG, water-soluble and oil soluble (sometimes referred to as 'insoluble') [29]. Both provide excellent film strength, viscosity, and temperature tolerances. More commonly used is water-soluble PAG (for purposes of ease of manufacture as well as biodegradability) that has general applications that are usually reserved for extremely high temperatures; some examples are heating fluid (for a heat pump) or bearing grease, as well as quenching fluid and food-grade lubrication. The major drawback that occurs with water-soluble PAG is that it reacts poorly with petroleum-based oils, and thus equipment must be completely overhauled in order to use PAG as a lubricant. This problem does not exist with synthetic esters, however. The major benefits PAG has over synthetic esters presents itself in the form of biodegradability; while synthetic esters are excellent lubricants that are easy to manufacture, the higher biodegradability versions offer less service time, whereas PAG has similar lubrication and wear properties but is easy to manufacture as both biodegradable as well as very stable.

PAG oil also continually offers excellent wear factors and offers high stability even at the end-of-life of the oil. The only factor that is affected is that of acidity, which can easily be worked away almost entirely with an additive package, and never reaches toxic levels [29]. Viscosity continues to be good even after extreme exposure to heat, oxygen, and water. PAG oils exceed synthetic esters in terms of water allowances, as well as non-environmentally friendly lubricants, especially when the PAG is water soluble, with water insoluble PAG oil still existing with a high-water contamination tolerance. The list of PAG oil uses continues, with that of industrial lubricant that is not changed often, or where maintenance might be unfeasible, such as a backhoe or excavator that is located far from civilization (a mine, for instance). This is, of course, due to PAG oils' heat, water, and oxidation resistance. PAG oil use is highly recommended in humid and moisture-rich climates. Finally, PAG oils' excellent wear factor can be used in the creation of parts and tooling, as well as any application where micro-pitting is an issue, such as any metal-to-metal contact with high loads and low velocity [29].

In terms of the ability of grease to attract or repel dirt or dust, the environmentally friendly grease performs similarly to the conventional grease. Both types of greases attract a significant number of contaminants. While not exactly something that it is heavily criticized for, it is interesting to note that the solutions to overcome the contaminants issue for environmentally friendly greases are similar to those for lithium-based grease in any weather or temperature. The solutions for both are quite simple, wipe away then reapply as necessary.

Next is their corrosiveness. Lithium grease is non-corrosive, which is desirable, as that is one of the reasons why one would apply grease to metal or plastic parts. The chemical makeup of all sorts of lithium grease is stable, and so it will not react to whatever it comes into contact with [30]. Biodegradable greases use vegetable oil, which consists of fatty acids that are also acidic. This means that the parts that the biodegradable grease gets applied on can corrode. However, a study done by Abdulbari [31] has shown that biodegradable greases are only slightly corrosive. This study did a copper corrosion test, and the result came back as 1b (dark orange), which means there is a slight tarnish. The copper corrosion test uses discoloration or a change of color to a copper strip to determine corrosivity, with 1a being freshly polished and a copper that is not corroded, and 4c being a black, highly corroded copper strip. In another study, fatty acid molecules (octadectrienoic acid and hexadecanoic acid) have shown corrosion inhibition of mild steel due to the presence of active functional groups with capability of fast electron transfer [32].

Next is their dropping point. A dropping point in grease is the highest point in temperature in which grease turns from a semi-solid substance to a liquid substance. This is very important as we do not want the applied grease to break down or leak out from where it was originally, which would lead to increased friction on the parts that it was applied on. For the conventional grease, there are two different brands that we will be looking at. One is from a control group that was used by a study done by Shapovalov *et al* [33] as a different perspective or reference point. Shapovalov *et al* developed rape oil-based grease using different amounts of soap thickener, sodium liquid glass and acetic acid. Litol-24 was used as a control grease which, from the company's technical data sheet, has a drop point of around 185 °C. This rape oil-based grease with 20, 5 and 2.8 wt.% of soap thickener, liquid glass and acetic acid yielded the drop point of 180 °C which is nearly similar to that of Litol-24, whereas this formulated grease showed an exceptional critical seizure load value of 680 N compared to 600 N of the Litol-24.

Another important aspect to compare them to is their coefficient of friction. The same study done by Shapovalov *et al* [33] also conducted a test on both of the greases' coefficient of friction. It has been found that the environmentally friendly grease performed worse than commercial grease in terms of coefficient of friction, such that the eco-grease has a higher coefficient of friction in all four compositions that they have tried.

Another study done by Sánchez *et al* [34] reported that castor oil as their base for the biodegradable grease yielded less or similar friction coefficient in comparison with lithium grease. The performance of the castor oil-based grease strongly depended on the thickener agent and tribological contact conditions.

Performance-wise, current biodegradable grease technology still needs improvement to match the performance of conventional greases. The only one that performed like a conventional grease is the dirt or contamination test. Biodegradable grease is slightly more corrosive than lithium-based grease. Economically friendly greases functioned nearly as well, but not at the same level as conventional greases, with the drop point test, especially when another element is added to its composition. The only functions where the green grease did better than the lithium-based grease were its ability to be environmentally friendly and its cost or price in the market. Therefore, if the moving parts that are rubbing against one another do not require high-performance lubrication or are not too important, then the green grease is worth the reduced performance cost.

6.6 Market analysis of PAG and synthetic esters

The PAG oil has had incredible prolificacy, accounting for ~30% of the whole synthetic lubricant market [35], it is slowly being phased out. Synthetic esters only make up ~17% of the market [35]. While PAG oils have been used for many decades and have excellent stability, modern synthetic esters are becoming the go-to synthetic lubricant, with excellent lubricity, resistant to breakdown and oxidation, water-in-oil tolerance, and modern applications requiring incredible levels of thermal resistance. As the engines available to the citizenry improve in performance, better lubricants are needed to combat the high wear and heat factors. Space travel is gaining traction, finally, and the extreme temperature tolerances once again require a very modular lubricant, ideally at low cost and environmentally friendly, in the case of any losses. All of these factors along with many more are contributing heavily to the synthetic ester becoming the majority synthetic lubricant used worldwide. In 2013, the synthetic ester market was based almost halfway on compressor oils specifically, to the tune of 40.7% [36]. At that time, synthetic lubricants were still relatively reserved, with the majority of lubricants being composed of modified hydrocarbons. The synthetic ester market size was calculated to be $1.3 billion USD in 2013 [36]. Since then, the entire market size of synthetic lubricants has grown vastly and was estimated to be $32.2 billion USD in 2018, with a compound annual growth rate (CAGR) of at least 3%. Synthetic esters grew from few applications to a great variance, with hydraulic oils, metalworking fluid, and automotive engine oils growing in large proportion. Synthetic esters are growing in popularity mainly for their reversible manufacturing processes and biodegradability, as well as high-temperature stability and high flash point.

These facts are growing the synthetic ester market to higher and higher proportions, with the CAGR growing from 3% in 2013 to 6.4% as of 2020 [36]. The global revenue forecast was also undervalued, with the market size growing to $2.01 billion USD as of 2020, as opposed to the previous number of $1.3 billion USD in 2013.

While the synthetic ester market is growing at a fast pace, the demand for PAG oils is increasing as well, at a seemingly faster rate. Given the long-standing reputation and consistent improvement over many decades of use, PAG oils are being used for new and more various applications than in previous decades, with the American market and Asian markets both beginning to use PAG oils in the production of polyurethane foam for construction, aviation, automotive, and many other markets. Currently, the three largest uses for PAG oils are Polyurethane foam (28%), functional fluids (coolant, hydraulics), and lubricants [37]. With the new and quickly growing expanding foam market and usage, PAG is regaining any market that it was losing to other environmentally friendly lubricants, including synthetic esters. With its various uses, there is little question as to the current market forecast of $18 billion USD by 2026 [37], which would easily make up half of the entire synthetic lubricants market, which would grow proportionally. While the CAGR of synthetic esters was a respectable 6.4% [36], the PAG CAGR percentage is 9.5% [37], half higher again than that of the ester market. This is more than likely due to the various uses of PAG as well as the long-standing history and familiarity; while synthetic esters are environmentally friendly and non-toxic as well, PAGs see many applications outside of pure machinery work. PAGs can be used in food packaging, surfactants, and pharmaceuticals.

Further, the PAG market can be broken into different segments and analyzed; type, solubility, and application, to name a few. Mainly, three mainstay types of glycols are PAG, polyethylene glycol (PEG), and the more popular polypropylene glycol (PPG). PPG currently holds 20% of the PAG market for its uses as spray foam and insulation for construction. Its non-toxicity allows it to be used in food and medicinal storage and varying purpose plastics. Also, it is water-soluble, making an already biodegradable substance even more environmentally friendly. PEGs have excellent energy efficiency and viscosity ratings as well, which cement their use in multiple industries and further drive the PAG market to higher highs. By application, PEGs also have the most varied uses. Solvents, thickeners, and emulsifiers are all combined with PEGs to make cosmetics and moisturizers, a vast and expensive market that only continues to grow exponentially.

From this general market analysis, a few points can be inferred. Chiefly, PAGs have held the larger market advantage over other synthetic and biodegradable substances, including synthetic esters. While the market does indeed support the competition, and synthetic esters are an excellent extreme situation lubricant, they unfortunately do not have any other current uses other than that of a lubricant, and still require some additives in order to resist breakdown. PAGs have been in the market for half a century at this point and have shown themselves to be an effective and lasting product, with many uses being discovered consistently. PAGs also have more than one distinct grouping and are applied in a myriad of uses from pharmaceuticals, food grade plastics, cosmetics, and construction to also being able to be used as extremely stable lubricants with excellent biodegradability. With the correct additive package, and for most applications, PAGs are the seemingly better choice, especially where extreme temperatures will not be applicable. With the large CAGR, it is possible that the PAG market at some point obviates the synthetic ester market for compressor oils and machine lubricants in conjunction with the previously listed market segments.

Synthetic esters do have the edge if the space market continues to grow, and especially if the manufacturing process becomes more affordable, increasing production and eventual usage. Synthetic esters would also benefit greatly from more research and development, akin to the somewhat recent discovery that PAGs (specifically PPG) can be used for more than just lubricants. With synthetic esters' excellent viscosity and temperature attributes, they could gain a major foothold in the plastics market and usurp PAGs, especially for processes such as the creation of newer synthetic polymers.

6.7 Conclusions

Grease is a widely used lubricant material, particularly where a thick lubricant is required. Grease in the market today is primarily composed of refined mineral oil. Typically, it is processed in a manner that increases its viscosity to an extremely high level. The unfortunate need to find better solutions for alternative greases arises from the fact that mineral oil is toxic to humans among others. 'Environmentally friendly' grease is inherently difficult to achieve as grease needs to be waterproof as well as resistant to elemental and thermal breakdown. In this chapter, the synthesis of environmentally friendly greases, their types and their application are explored and compared with conventional greases. This chapter also provides an understanding of the benefits of using environmentally friendly grease and how grease itself could inherently be 'environmentally friendly.' Understanding the types of environmentally friendly grease that exist and that is in development is crucial in analyzing the legitimacy and feasibility of environmental grease.

References

[1] Morales-Espejel G E, Lugt P M, Pasaribu H R and Cen H 2014 Film thickness in grease lubricated slow rotating rolling bearings *Tribol. Int.* **74** 7–19

[2] Wang Z, Xia Y and Liu Z 2015 The rheological and tribological properties of calcium sulfonate complex greases *Friction* **3** 28–35

[3] Florea O and Luca M 2010 Multipurpose lubricating greases for food-grade applications *J. Balk. Tribol. Assoc* **16** 233

[4] Grives P R 2000 The manufacture of biodegradable nontoxic lubricating greases *NLGI Spokesman* **63** 25–9

[5] Florea O, Luca M and Steliean C 2005 Ecological lubricating greases *Environment* **5** 7

[6] Fan X, Wang L and Xia Y 2015 Oil-soluble lithium salts as novel lubricant additives towards improving conductivity and tribological performance of bentone grease *Lubr. Sci.* **27** 359–68

[7] Wang Z, Xia Y and Liu Z 2011 Study the sensitivity of solid lubricating additives to attapulgite clay base grease *Tribol. Lett.* **42** 141–8

[8] Martín Alfonso J, Yañez R, Valencia C, Franco J and Díaz M 2009 Optimization of the methylation conditions of Kraft cellulose pulp for its use as a thickener agent in biodegradable lubricating greases *Ind. Eng. Chem. Res.* **48** 6765–71

[9] Nagendramma P and Kumar P 2015 Eco-friendly multipurpose lubricating greases from vegetable residual oils *Lubricants* **3** 628–36

[10] Condat Corporation 2022 A complete range of biodegradable greases intended for the lubrication of all types of mechanism https://www.condatcorp.com/product/greases/biodegradable-grease/ (accessed on 21 January)

[11] Sander J 2022 Can lubricants be green? *BiomassMagazine* https://biomassmagazine.com/articles/17390/can-lubricants-be-green (accessed on 21 January)

[12] Meng Q 2017 The impacts of fracking on the environment: a total environmental study paradigm *Sci. Total Environ.* **580** 953–7

[13] OECD 1994 *OECD Guidelines for the Testing of Chemicals* Organization for Economic Development

[14] ASTM 2011 ASTM 5864: standard test method for determining aerobic aquatic biodegradation of lubricants or their components *Annual Book of ASTM Standards* (West Conshohocken: ASTM International)

[15] Fuels 1993 CECftDoPTfLaE. Biodegradability of two-stroke cycle outboard engine oil in water CEC-L33-A93
[16] Axelman J, Broman D, Näf C and Pettersen H 1995 Compound dependence of the relationship log K_{ow} and log BCFL *Environ. Sci. Pollut. Res.* **2** 33–6
[17] Hemmelskamp J and Brockmann K L 1997 Environmental labels—the German 'blue angel' *Futures* **29** 67–76
[18] Thidell Å, Leire C and Lindhqvist T 2015 *The Nordic Swan 2015: Performance Indicators for Ecolabelling* (Copenhagen: Nordic Council of Ministers)
[19] Gertz R 2005 Eco-labelling—a case for deregulation? *Law, Probab. Risk* **4** 127–41
[20] EN Official Journal of the European Union 2018 L 285/82 COMMISSION DECISION (EU) 2018/1702 of 8 November 2018 establishing the EU Ecolabel criteria for lubricants (notified under document C(2018) 7125) https://www.ecolabel.dk/-/criteriadoc/3790
[21] OSPAR. Oil Spill Prevention, Administration and Response (OSPAR) http://www.dem.ri.gov/programs/emergencyresponse/ospar-reports.php (accessed on 18 December)
[22] Adelman M A 1990 Mineral depletion, with special reference to petroleum *Rev. Econ. Stat.* **72** 1–10
[23] Nowak P, Kucharska K and Kamiński M 2019 Ecological and health effects of lubricant oils emitted into the environment *Int. J. Environ. Res. Public Health* **16** 3002
[24] Bartz W J 1998 Lubricants and the environment *Tribol. Int.* **31** 35–47
[25] Environmentally Acceptable Lubricants. United States Environmental Protection Agency, Office of Wastewater management. https://www3.epa.gov/npdes/pubs/vgp_environmentally_acceptable_lubricants.pdf (accessed on 19 Jun 2022)
[26] Aluyor E O and Ori-Jesu M 2009 Biodegradation of mineral oils–a review *Afr. J. Biotechnol.* **8** 915–20
[27] Greaves M 2020 Polyalkylene glycols *Synthetics, Mineral Oils, and Bio-Based Lubricants* (Boca Raton, FL: CRC Press) pp 119–46
[28] Boyde S 2000 Hydrolytic stability of synthetic ester lubricants *J. Synth. Lubr.* **16** 297–312
[29] Beatty D and Greaves M 2006 Polyalkylene glycol synthetic PAG oil explained *Machin. Lubric.* p 23
[30] Fan X, Xia Y, Wang L, Pu J, Chen T and Zhang H 2014 Study of the conductivity and tribological performance of ionic liquid and lithium greases *Tribol. Lett.* **53** 281–91
[31] Abdulbari H A, Zuhan N and Mahmood W K 2017 Biodegradable grease from palm oil industry wastes *J. Adv. Res. Fluid Flow* http://uest.ntua.gr/athens2017/proceedings/pdfs/Athens2017_Abdulbari_Mahmood_Yunus.pdf
[32] Khanra A, Srivastava M, Rai M P and Prakash R 2018 Application of unsaturated fatty acid molecules derived from microalgae toward mild steel corrosion inhibition in HCl solution: a novel approach for metal–inhibitor association *ACS Omega* **3** 12369–82
[33] Shapovalov V, Zlotnikov I, Kupchinov B, Timoshenko V and Zubritskii M 2007 New biodegradable antifriction grease *J. Frict. Wear* **28** 306–9
[34] Sánchez R, Fiedler M, Kuhn E and Franco J 2011 Tribological characterization of green lubricating greases formulated with castor oil and different biogenic thickener agents: a comparative experimental study *Ind. Lubr. Tribol.* **63** 446–52
[35] MRFR research 2022 *Polyalkylene Glycol Market Size, Share, Growth | Report, 2030*
[36] GrandViewInc. 2022 *Synthetic Ester Lubricants Market Size & Share Report, 2020*
[37] iCrowdNewswire 2021 *Polyalkylene Glycol Market Size Forecast to Reach $18 Billion by 2026*

IOP Publishing

Multifunctional Bio-Based Lubricants
Synthesis, properties and applications
Ashish K Kasar, Arpith Siddaiah and Pradeep L Menezes

Chapter 7

Additives for lubricants

Ashish K Kasar, Logan Richardson, Alexander Scherp, Chris Tyree, Jacob Welty and Pradeep L Menezes

This chapter aims to examine the performances of various additives in lubricants and the classification of these additives. In this chapter, the lubricant additives were classified based on their function, i.e., surface protection, performance, lubricant performance, and multifunctional nature. The various additives were examined for their performance in their designated use. The performance of these additives was critically evaluated by reviewing the major studies containing experiments and results. In addition to the individual functionality of the additives, the multi-functionality of suitable additives was highlighted. The environmental suitability of these additives was also discussed.

7.1 Introduction

Surface protective additives play a major role in the performance of lubricants when in contact with structured surfaces. There are a variety of surface protective additives that keep the lubricant and surface in performing shape. These additives include anti-wear agents, corrosion/rust agents, detergents, and dispersants. Without these protective additives, the lubricant will be a solution for friction, but the surface will be susceptible to other failures. Also, adding too many protective additives will not benefit the oil but may cause its performance to deteriorate. The addition of one additive in the material may increase that additive-resistant property but may harm another. When blending additives with oil, all additives must be balanced to achieve a dynamic solution.

Performance additives to lubrication play a key role in ensuring the longevity, performance, and efficiency of various mechanical systems and oils. In this chapter, performance additives are categorized by pour point depressants, seal swell agents, viscosity improvers, extreme pressure additives, and friction modifiers. Under performance additives, the analysis and discussions were made on the viscosity improvers, pour point depressants, and friction modifiers.

doi:10.1088/978-0-7503-3435-8ch7

The lubricant protective additives consist of anti-foaming agents, anti-oxidants, and metal deactivators. Anti-foaming agents serve the purpose of preventing the formation of foam and decreasing the stability of any foam that does form by reducing the surface tension. Anti-oxidants increase the time it takes for a lubricant to oxidize and break down by decreasing or eliminating the presence of free radicals in a lubricant. Metal Deactivators are similar to anti-oxidants in the sense that the latter help to increase the time it takes for a lubricant to oxidize, but the former (metal deactivators) work by forming a film on metal surfaces and reducing small metallic particles and their effect on the oxidation in a lubricant. Metal deactivators and anti-oxidants often complement each other very well and therefore are most commonly seen working together in lubricants.

Multifunctional additives are additives that serve more than one purpose when added to the base oil. These additives can improve multiple functions of a lubricant such as anti-wear, extreme pressure, anti-rust, viscosity index, antioxidant, anti-friction, and more. The study of multifunctional additives can be beneficial in decreasing the wear and corrosion of moving parts. The additives are often tested and compared to a base oil or other lubricant additives to examine their tribological performance. They can also be used to reduce the environmental impacts of the additives used today that are not biofriendly.

7.2 Classification of additives based on functionality

7.2.1 Surface protective additives

7.2.1.1 Anti-wear agents

Anti-wear additives are used to prevent metal-to-metal contact. These additives chemically react with the metal surfaces by forming a lubricious coating that protects the metal from wear under boundary lubrication conditions. Anti-wear additives produce a harder and more protected surface than the ordinary base metal. These additives are most used in formulated engine oils that have been used in automotive, industrial, and turbine engines. The most generic anti-wear agent used in commercial oil today is zinc dialkyldithiophosphate, better known as ZDDP [1]. ZDDP has been used as an analytical component to all modern automotive applicable engine lubricants for over 60 years. It limits wear primarily by forming a thick, protective, phosphate-based tribofilm on rubbing surfaces. The film creation can occur at low temperatures (26 °C) and is relatively unchanged from the chemical difference of the base substrate [1]. The tribofilm consists of primarily amorphous zinc phosphate and is relatively rough and thick (200 nm). Figure 7.1 shows an AFM topography map and profile of a developed ZDDP tribofilm. The profile shows the thickness separated by deep valleys, which help ZDDP limit the amount of direct contact between the two rubbing surfaces [1].

At high temperatures, usually above 150 °C, ZDDP tribofilm forms more slowly when two surfaces rub together. The film is called 'thermal films' and has different characteristics than standard ZDDP tribofilm. At these temperatures, ZDDP is known for causing deposits upon thermal decomposition known as 'ash'. The thermal deposits are the result of the zinc concentration in the additive.

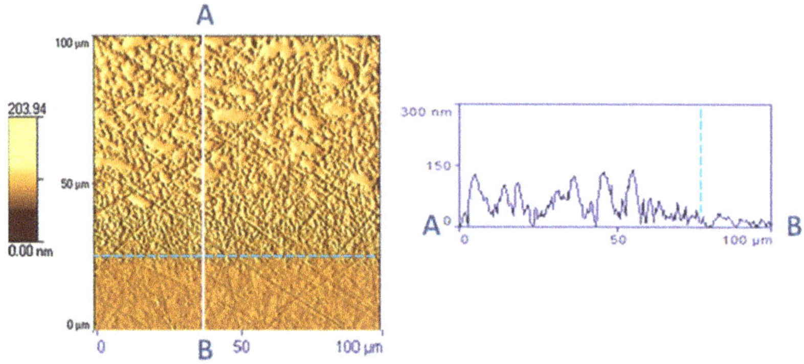

Figure 7.1. ZDDP tribofilm AFM topography and profile. Reprinted by permission from *Tribology Letters* Springer Nature [1], copyright 2016.

Qu *et al* [2] explored oil-miscible ionic liquid where ZDDP was used as an anti-wear agent. The authors explored the anti-scuffing and anti-wear behavior and mechanics of an ionic liquid (IL) trihexyltetradecylphosphonium bis(2-ethylhexyl) phosphate. Ionic liquids have been explored as ashless lubricant additives due to their excellent thermal stability, low-vapor pressure, and strong wear protection in the tribofilm boundary-layer formation [2]. The main technical barrier with ILs is that they have no or very little solubility in non-polar hydrocarbon base oils. This is due to the ions being attracted to polar forces, and even hydrogen bonding. In non-polar hydrocarbon oils, the molecules are held together by dispersion forces (Van Der Waals). Qu *et al* [2] developed oil-miscible ILs (phosphate-based) by designing large three-dimensional ionic molecules to dilute the charge and thus improve the solubility with neutral oil molecules. Also, these phosphate ILs are non-corrosive, thermally stable, and have shown strong anti-scuffing and anti-wear functionality.

In the same study, authors compared trihexyltetradecylphosphonium bis(2-ethylhexyl) phosphate and ZDDP as anti-wear/anti-scuff additives for Mobil 1™ poly-alpha-olefin (POA) (Exxon Mobil Corp.). Both the IL and ZDDP were in POA at 1.0 wt.%. The calculated phosphorus levels for IL-POA and ZDDP-POA are 772 and 1000 ppm, respectively. The authors conducted a tribological reciprocating sliding test on a Plint TE77 high-frequency tribometer. This evaluation replicates engine piston top rings sliding with grey cast iron cylinders from General Motors engines.

The IL-POA and the ZDDP were evaluated at both room temperature and 100 °C for the reciprocating sliding test. Both additives performed effectively in preventing scuffing at room temperature. However, at 100 °C the base oil viscosity was much lower and ZDDP was unable to prevent scuffing damage and caused deposits due to thermal decomposition. The IL additive at 100 °C maintained stable boundary lubrication and low wear rate throughout the test. The study suggests that the IL additive possesses excellent anti-wear characteristics compared to conventional ZDDP. This potentially enables the use of lower viscosity oils for engine lubrication, thus increasing fuel economy.

7.2.1.2 Corrosion inhibitors

Corrosion inhibitors are chemical compounds that slow and prevent the corrosion rate of a material, typically metal or an alloy that encounters fluids. Corrosion is an electrochemical process in which metal (or alloy) reacts in the presence of an electrolyte. Corrosion inhibitors function by absorbing onto the metal surface, which then blocks the surface of the metal from the effects of water, air, and acids. Corrosion in the presence of aggressive chemicals such as acids and bases requires different additives for protection. The chemical corrosion inhibitors form an inactive film on the metal surface by interacting with metallic ions that reside on the surface. Some chemical corrosion inhibitors work by neutralizing corrosive acids that form from the oil and additive degradation.

Polyaniline (PANI) has been evaluated as an ideal material due to its excellent environmental stability, tunable conductivity, controllable structure, and novel anti-corrosive properties [3]. Additionally, PANI is formulated with simple chemical or electrochemical polymerization, which allows for the control of nanostructures through reaction parameters. PANI has also exhibited friction reduction and anti-wear abilities under various conditions when evaluated in lubricating grease. Cao *et al* [3] evaluated ILs, which are organic salts that have remarkable thermal stability, low volatility, and negligible vapor pressure. Imidazolium-based ILs are a certain type of ILs that have been found to have hydrogen bonding interactions with PANI. This led to the usage of PANI with imidazolium-based ILs, which could exhibit excellent anti-corrosive properties, as well as tribological properties.

The PANI doped with ILs in poly alpha olefin (PAO) was assessed for tribological and corrosive tests. The friction/wear test was conducted on an MFT-R4000 reciprocal friction and wear tester with a ball-on-disc (steel/steel). The upper ball (52100 steel, diameter 5 mm) slides reciprocally against a fixed lower disc (52100 steel) at a frequency of 5 Hz, a normal load of 50–200 N at room temperature. This study examined the tribological characteristics of two types of IL-doped PANI, which are PANI-LP204 (rod-shaped) and PANI-LB104 (irregular shape). Figure 7.2 shows the observed coefficient of friction (COF) and wear volumes based on additive concentration for both types of IL-doped PANI.

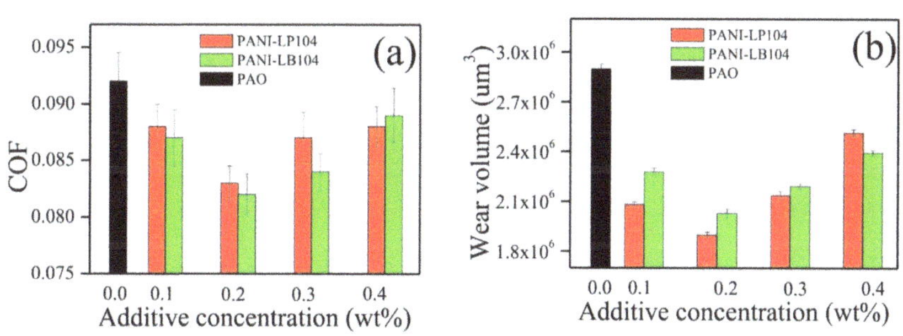

Figure 7.2. (a) COF and (b) wear volumes for PAO with different concentrations of IL-doped PANI. Reprinted from [3], copyright (2018), with permission from Elsevier.

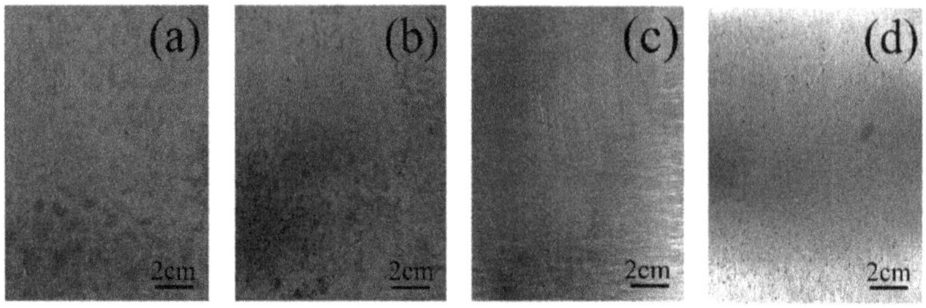

Figure 7.3. Steel block coated with Basic Polyurea Grease (a) and (b) and IL-doped PANI (c) and (d) after the corrosion salt spray test. Reprinted from [3], copyright (2018), with permission from Elsevier.

The anti-corrosion characteristics acquired from the PANI were evaluated with a salt spray test. Bright finished steel blocks were evenly coated with both types of IL-doped PANI and pure polyurea grease as a control. The blocks were then placed into a salt spray test chamber at 35 °C and a NaCl solution covered the steel blocks for 720 h. The blocks were then washed with acetone and air-dried and photographed for examination. The results of the test were evaluated on a PHI-5702 multifunctional x-ray photoelectron spectroscopy and examined the binding energy of the different additives. Figure 7.3 shows the corrosion test results, with a noticeable difference between the blocks coated with the IL-doped PANI and conventional polyurea grease. As shown, the IL-doped PANI (PANI-LP104 and PANI-LB104) exhibit excellent anti-corrosion properties.

Based on the tribological friction/wear test and the salt spray corrosion test performed, the IL-doped PANI exhibited excellent anti-wear and anti-corrosion properties. Additionally, due to the simple and synthetic approach of combining the additives, IL-doped PANI is an ideal additive that holds great potential in the lubrication domain.

7.2.1.3 Detergents

Detergents play an important role in several tribological applications but are primarily used in engine oils. Detergents help to keep hot metal components free of deposits (clean) and neutralize acids and other chemicals that form in the oil. They are capable of neutralizing acids due to them being naturally alkaline or basic. Detergents form the base of alkalinity in engine oils, which are referred to as a base number (BN). Detergents usually consist of calcium and magnesium chemistry, but barium-based detergents are also available but rare.

Over-based calcium detergents are well known for being excellent anti-wear additives, as well as having their inherent detergent characteristics. Over-based calcium alkylbenzene sulfonate (OCABS) is an over-based detergent and a potential tribological additive [4]. The evaluation of OCABS is in response to improving engine oil additives to increase the overall performance of the engine by minimizing deposits, corrosion, and varnish in engine hot areas.

Tribological evaluations were performed to analyze OCABS surface wear/friction properties with a Cameron-Plint tribometer in a cylinder-on-flat configuration to replicate an engine environment [4]. OCABS is added to Poly-Alfa-Olefin (PAO) at a concentration of 3 wt.%, and all evaluations were performed in UHV (ultra-high vacuum) conditions. The examinations are performed in UHV conditions because impurities or absorbed particles do not pollute friction surfaces in these conditions, which allows for constant friction surface characterization. The cylinder-on-flat rubbing surfaces were made of AISI 52100 polished steel and run for 1 h with a mean contact pressure of 850 MPa. The mean contact force of the cylinder is 350 N and runs with a sliding speed of 200 mm s^{-1} at a stroke length of 7 mm. The test temperature is about 25 °C and rises to 40 °C due to friction, by the end of the experiments. The results show that PAO with 3 wt.% OCABS outperformed conventional pure PAO and shows excellent anti-wear/scuff properties.

Additionally, the friction coefficient for the PAO with 3 wt.% OCABS was significantly less than pure base oil of PAO, especially at the initial period of applying load that suggests the superb anti-wear/scuff and inherent detergent abilities of the OCABS.

Cizaire *et al* [4] concluded that due to the several advantages that OCABS possesses, over-based detergents act as excellent anti-wear additives while being able to prevent residue and clean rubbing surfaces.

7.2.1.4 Dispersants

Dispersant's main function is to keep particles of soot developed from rubbing surfaces finely dispersed and spread throughout the oil (less than 1 μm in size). The objective is to keep the containment distributed and not allow the soot to group up and cause damage to the lubricant or surface. Dispersants are generally organic and ashless, thus being not easily detectable with conventional tribological analysis. The combination of detergent/dispersant additives allows for more acid compounds to be neutralized and the soot byproduct to stay dispersed.

Engine soot is created by the high wear/temperature decomposition of base oils and lubrication additives such as ZDDP. High levels of soot have been directly reported to induce higher levels of wear in engine components. There have been several theories involving that certain anti-wear additives not only create excess soot but absorb it, changing their anti-scuff properties. ZDDP and common engine surrogate soot carbon black (CB) have been examined together and shown to increase pin-wear through a tribological pin test [5]. ZDDP is the most used conventional anti-wear in gasoline and diesel engines due to its long time of use in the industry, and ability to create an efficient tribological film on the rubbing material's surface. Further study is implicating that there is a need to improve upon this ZDDP or add an extra additive to improve the performance of ZDDP. Kontou *et al* [5] performed wear tests with a high-frequency reciprocating rig (HFRR) controlled with a ball-on-flat friction/wear tester with three different ZDDP zinc blends, as well as eight different common PIBSA polyamine dispersants. The reason was to examine a large quantity of specimens for thorough tribological results. This also examines the different blends of ZDDP based on primary/secondary zinc levels, which affects the performance of the additive due to thermal decomposition. The evaluations were tested with a range of different dispersant concentrations

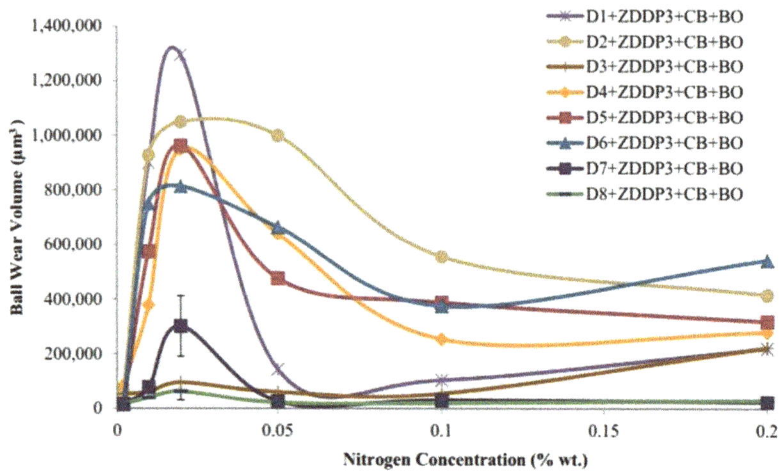

Figure 7.4. Wear volume versus nitrogen concentration for ZDDP-3 and different dispersants [5].

(0.002, 0.02, 0.05, 0.1, 0.2 wt.%). The tester was set and engine environment conditions with a 3.92 N load, 1.03 GPa contact pressure, 100 °C temperature, at a frequency of 50 Hz for 60 m.

Additionally, a rheology test was conducted to test the dispersant ability for CB and was sampled with different nitrogen concentrations. These examinations were carried out on a Peltier plate with a conical geometry of 40 mm at a 0.5° cone angle, and a controlled temperature of 100 °C. HFRR ball wear volume tests had shown the best results with ZDDP-3, which is the secondary zinc blend. Figure 7.4 shows the test results based on wear volume (anti-wear) and nitrogen concentration (detergent efficiency).

As shown in figure 7.4, wear volume significantly varied with different dispersants, and proved that there are proficient dispersants for improving anti-wear properties. The low level of volume wear from dispersant eight proves that high concentration levels of polyamine dispersants appear to protect the iron-sulfide file from abrasion and prevent high levels of wear for ZDDP and excessive soot [5].

Each of these additives plays a successful role in protecting the contact surface of objects in tribological applications. A list of various other surface protective additives is provided in table 7.1.

7.2.2 Performance additives

Performance additives play a critical role in protecting mechanical parts inside and outside, preventing premature wear and aging, and changing tribological properties based on the desired criteria. Performance additives aim to achieve these three goals along with a variety of others. In this section, various performance additives to base oils and their effect on tribological properties are discussed.

Table 7.1. Various surface protective additives.

Additives	Base oil	Wt.%	Testing conditions, material, load, velocity	Tribology properties	Purpose	References
Trihexyltetradecyl-phosphonim bis(2-ethylhexyl)phosphate	Poly-alpha-olefin 4-cSt (Mobil 1)	1.0	Plasma coating on the grey cast iron sprayed, reciprocating sliding test, at 160 N and 240 N, 10 Hz	Thermally stable due to low zinc levels	Anti-wear	[2]
Zinc dialkyldithiophosphate (generic anti-wear additive)	Poly-alpha-olefin 4-cSt (Mobil 1)	1.0	Reciprocating sliding test Plint TE77, grey cast iron sprayed w/a plasma coating, 160 N and 240 N, 10 Hz.	Not thermally stable, leaves 20% soot at high temp (breaks down), high phosphorus	Anti-wear	[2]
Ionic liquid doped polyaniline (PANI)	Poly-alpha-olefin and polyurea grease	Not identified	MFT-R4000 reciprocal and wear tester (ball-on-disc), steel to steel, 5 Hz, 50–200 NCorrosion test: salt spray test	'Remarkably reduce friction coefficients'Anti-wear properties, environmentally stable, conductive	Corrosion agent (IL doped)	[3]
Calcium alkylbenzene sulfonate (OCABS)	Poly-alpha-olefin (POA)	3.0	UHV conditions, Cameron-Plint tribometer, cylinder-on flat, AISI 52100 polished, P_{MEAN}= 850 MPa, F_{NORM} = 350 N	No significant effect of anti-wear, does not need to be thermally activated	Detergent	[4]
PIBSA polyamine dispersant (mixed w/ZDDP)	Gas to liquid (GTL) base oil with viscosity 18.3 cSt at 40 °C and 4.2 cSt at 100 °C	0.4–1.0	HFRR wear tester, controlled reciprocating friction and wear testing, ball and disc, 1.03 GPa CP, 50 Hz, 3.92 N, and 100 °C	Disperses well in high phosphorus ZDDP, and allows for the dispersion of carbon black (zinc soot surrogate) allowing for a higher quality ZDDP	Dispersant	[5]
Cashew nut shell liquid (CNSL)	API P110	0.280, 1.220, 0.016, 0.002, 0.01	Corrosion analysis and gravimetric analysis (300 mL autoclave from 298 K to 353 K for 24 h)	Formulations at 333 K showed higher efficiency of corrosion resistance with CNSL as well as preventing surface damage	Corrosion Agent	[6]

7.2.2.1 Friction modifiers

There is a growing need for energy efficiency. A direct result of this necessity is an increased need for friction modifiers. The most common friction modifiers are functionalized polymers, soluble organo-molybdenum, and organic friction modifiers. A few types of organic friction modifiers are stearic acid, oleic acid, and hexadecane additives. Secondly, oil-soluble organo-molybdenum friction modifiers were first patented as an additive to lubricants in 1939 [7]. The two main types of oil-soluble molybdenum friction modifiers are molybdenum dialkyldithiophosphate (MoDDP), molybdenum dialkyldithiocarbamate (MoDTC), and functionalized polymers.

Through years of research and analyzing the four friction modifier additives, the researchers were able to conclude that organo-molybdenum additives in the presence of rubbing asperities create MoS_2 nanosheets which have low shear strength properties. This is beneficial in terms of friction modifiers because these nanosheets will separate the rubbing asperities and overall will decrease friction in various systems and applications. Functionalized polymers were found to postpone the inevitable mixture lubrication by absorbing and forming high viscosity surface films. These surface films allow surfaces to not be in full contact in the presence of low entrainment speed conditions. The functionalized polymer has potential in a variety of mechanical systems where friction due to rubbing surfaces negatively affects the overall function of the system.

7.2.2.2 Viscosity improvers

Mostly, viscosity improvers are polymers. Dandan *et al* [8] investigated the effect of polymeric viscosity improvers on palm kernel methyl ester (PKME). In this study, the polymeric viscosity improver is known as ethylene-vinyl acetate copolymer (EVA copolymer). The authors used vegetable oil which is chemically treated PK oil (PKME) and it was in semi-solid form at room temperature. The EVA copolymer has a vinyl acetate content of 25%, its appearance is colorless, its density at 25 °C is 0.948 g cm^{-3}, has a melt index of 19 g/10 min, and has a melting temperature of 75 °C. PKME was created using 200 mL of PK oil and a 6:1 methanol-to-oil ratio was used under 1.5 wt.% of calcium oxide acting as a catalyst. After the PKME had been created, it was blended with 2, 3, and 4 wt.% of EVA copolymer.

A four-ball tribotester machine was used to evaluate both the anti-wear and extreme pressure characteristics of each of the three lubricants. The ball bearings were AISI E-52100 chrome alloy with a diameter of 12.7 mm and an HRC hardness of 64–66 RC. The four-ball tribotester was performed under ASTM D2783 standards, which state that the testing conditions are at 35 °C, a speed of 1760 rpm, a duration of 10 s, and a starting load of 40 kg until failure.

The results show that the PKME oil had significant improvement from the base PK oil (vegetable oil) and continued to increase with the addition of the additive EVA copolymer. The results show that PK oil had a kinematic viscosity of 35.4 at 40 °C and 17.6 at 100 °C. PKME had a kinematic viscosity of 6.2 at 40 °C and 2.3 at 100 °C. The viscosity values increased with an increase in EVA content. PKME with 4 wt.% EVA copolymer had a kinematic viscosity of 43.7 at 40 °C and 15.4 at 100 °C.

The VI values were 238, 373, 406, and 368 for the PKME, PKME+2 wt.% EVA, PKME+3 wt.% EVA, and PKME+4 wt.% EVA, respectively. This demonstrates that the application of EVA copolymer to PKME oil significantly increases its kinematic viscosity.

7.2.2.3 Pour point depressants

The pour point of an oil is referred to as the temperature at which it can no longer flow due to the creation of crystalline nets, which trap the liquid hydrocarbon molecules, causing it to no longer be able to flow. Reduction of pour point is highly significant for the application of bio-based oils to replace the petroleum-based diesel fuel. Soriano *et al* [9] ozonized vegetable oils as a pour point depressant for biodiesel. Prior to ozonation, transesterification of sunflower, soybean, rapeseed, and palm oils was carried out to achieve methyl ester by removing glycerol. The resulting oil was ozonized by passing the ozone–oxygen mixture. These ozonized vegetable oils were added to the biodiesel in a small amount of 1–1.5 wt.%. The resulting pour point of biodiesel prepared from the ozonized sunflower oil, soybean oil, and rapeseed oil were −24 °C, −12 °C, and −30 °C, respectively, whereas the biodiesel with palm oil did not show any significant decrease in pour point. The improvement in the pour point by ozonized oil was observed due to the formation of smaller and regular shaped solids at low temperatures rather than the agglomeration of crystals.

Additives such as poly alpha olefin and diisodecyl adipate have also been studied to reduce the pour point of vegetable oils [10]. The addition of 1 wt.% of poly alpha olefin and diisodecyl adipate lowered the pour point of canola oil and sunflower oil to −33 °C and −24 °C. The study also suggested that further addition of the additives, i.e., dilution of oils did not lower the pour point. The most common pour point additive is poly alpha olefin which has been utilized in the waste cooking oil–biodiesel blend [11], and vegetable ester oils [12], and is also commonly added in the petroleum-based diesel fuels [13].

7.2.2.4 Anti-wear and anti-friction additives

When designing a mechanical system, friction and wear between mechanical parts must be taken into consideration. The results of this friction and wear are loss of energy, premature failing of parts, and a shortened lifespan of the given mechanical parts. To combat friction and wear, lubrication is key. In this section, various anti-wear and anti-friction additives are discussed. The solid lubricants are suitable candidates as anti-friction additives for vegetable oils. Solid lubricants such as graphene, hBN, graphite, and MoS_2. Reeves and Menezes [14] investigated the effect of hBN addition on the friction and wear properties of avocado and canola oil. The COF resulted in the presence of 70 nm hBN-avocado oil was ~64% less than that of no additive-avocado oil. The study revealed that the size of the solid lubricant additive plays a dominant role in controlling friction and wear. The COF resulting in the presence of 70 nm hBN-avocado oil was 60% less than that of 5 μm hBN-avocado oil. A similar reduction in wear volume was reported. The authors applied a similar methodology to check the efficiency of different sizes of hBN, MoS_2, carbon nanotube, WS_2, and graphite additives for the environmentally friendly phosphonium-based ionic liquids [15]. These additives can also reduce the wear by the formation of a tribofilm of liquid lubricant and solid lubricants. Additionally, few especially anti-wear additives are used, such as nano powders of metal oxides [16], carbides, and borates [17]. These hard particles are used for anti-wear properties under extreme pressure conditions. These nanoparticles form a physical protective film that reduces friction and wear. A list of various other performance additives is provided in table 7.2.

7.2.3 Lubricant protective additives

Lubricant protective additives serve the purpose of protecting the base lubricant. These additives consist of anti-foaming agents, anti-oxidants, and metal deactivators. Anti-foaming agents serve the purpose of preventing the formation of foam and decreasing the stability of any foam that does form by reducing the surface tension. Anti-oxidants increase the time it takes for a lubricant to oxidize and break down by decreasing or eliminating the presence of free radicals in a lubricant. Metal deactivators are similar to anti-oxidants in that they help to increase the time it takes for a lubricant to oxidize, but metal deactivators work by forming a film on metal surfaces and reducing small metallic particles and their effect on the oxidation in a lubricant. Metal deactivators and anti-oxidants often complement each other very well and therefore are most commonly seen working together in lubricants.

7.2.3.1 Anti-foaming

Foaming in lubricants is often caused by additives in lubricants such as anti-oxidants or detergents while at high flow, high pressure, or being penetrated by air and other particles. Foaming in lubricants brings about numerous problems such as loss of oil caused by oil overflow, high pressures, and inconsistent lubrication. These problems often cause increased heat, wear, and friction which also leads to increased cost, pollution, and reduced machine component life. In order to limit all these issues brought about by the foaming of lubricants, an anti-foaming agent is introduced to the lubricant. Anti-foaming agents help increase the resistance to foaming by decreasing the surface tension of the foaming lubricant to quickly break down the bubbles. The anti-foaming agents reduce the tendency of growth and the stability of the foam.

7.2.3.1.1 Polydimethylsiloxane (PDMS)

Polydimethylsiloxane is a silicone-based organic polymer that is used here as an anti-foaming agent. Prolic *et al* [22] tested PDMS for its effect on foaming and wear at 15–30 PPM in six different base lubricants. (1) Biodegradable synthetic-based hydraulic lubricant (HEES), (2) tractor oil (UTTO), (3) circulating oil, (4) synthetic transmission gear oil (SAE 75W-90), (5) hydrodynamic transmission oil, and (6) neat metalworking oil. For the foaming test, where the tendency of growth and the stability of the foam were tested, each base oil was tested at three different temperatures. Each sample was then penetrated by air for five minutes, and the volumes were recorded before and after to obtain the tendency of foaming.

Table 7.2. Performance additives for lubricants.

Additives	Base oil	Wt.%/volume	Purpose	References
Stearic acid	Hexadecane	Not identified	Friction modifier	[7]
Elaidic acid	Hexadecane	Not identified	Friction modifier	[7]
Oleic acid	Hexadecane	Not identified	Friction modifier	[7]
Ethylene-vinyl acetate copolymer (EVA)	Palm kernel methyl ester (PKME)	4% EVA copolymer was found to be ideal	Viscosity improvers. Test to show at what load welding occurs	[8]
Graphene	4-*n*-Pentyl-4′-cyanobiphyl Liquid Crystal (5CB)	5CB, +0.05 wt.%, +0.15 wt.%, +0.25 wt.%, +0.35 wt.%	To lower the wear and friction of mechanical parts	[18]
MoS_2 and SiO_2	EOT5# Engine oil	0.2 wt.%, 0.5 wt.%, 0.7 wt.%, 1.0 wt.%	Analyze the magnesium alloy/steel contact	[19]
Attapulgite powders (AP)	150 SN	SiO_2 (58.88 wt.%)MgO (12.1 wt.%) Al_2O_3 (9.5 wt.%)FE_2O_3 (5.2 wt.%) K_2O (1.04 wt.%)CaO (0.4 wt.%) TiO_2 (0.55 wt.%)P_2O_5 (0.18 wt.%) MnO (0.05 wt.%)Cr_2O_3 (0.04 wt.%)H_2O (12.06 wt.%)	To reduce the average COF and the wear rate of 150 SN with the various Attapulgite powders as an additive	[20]
PAO 4 oil	Diesel soot	1 wt.% sorbitan monooleate (Span 80)	To reduce the friction and wear of different soot concentrations	[21]

Table 7.3. Results of the foaming test due to the anti-foaming agent PDMS [22].

	Decrease in tendency with PDMS (mL)			Stability with PDMS (mL)		
Base oil	75 °C	93.5 °C	24 °C	75 °C	93.5 °C	24 °C
1	410	20	540	0	0	0
2	50	280	60	0	0	0
3	400	0	270	0	0	0
4	125	80	200	0	0	0
5	350	20	220	0	0	0
6	360	260	440	0	0	0

Table 7.4. Result of resistance to wear test due to the anti-foaming agent PDMS [22].

	Decrease in wear diameter with PDMS (mm, %)		
Base oil	1200 RPM	1800 RPM	2200 RPM
1	0, 0	0.02, 5	0.08, 18
2	0, 0	0.10, 23	0.07, 15
3	0.02, 3	0.03, 4	0.13, 5
4	0.01, 1	0.03, 4	0.04, 5
5	0.06, 10	0.02, 4	0.27, 30
6	0.01, 2	0.02, 3	0.14, 16

The stability of the foam was then measured following a 10 min waiting period. The results of the six base lubricants are shown in table 7.3.

Resistance to wear was tested next using a four-ball test apparatus. In addition, load-carrying capacity was evaluated. The test was controlled with a load of 392 N at 75 °C, for 1 h. The test was performed at three speeds (1200, 1800, and 2200 rpm). The results are shown in terms of decrease in wear diameter, mm, with PDMS in table 7.4.

Based on the results from these two tests, it is easy to note that polydimethylsiloxane (PDMS) works extremely effectively and provides anti-foaming properties to lubricants. PDMS greatly reduced the foaming tendency of all six base oils and at all three different temperatures as well as eliminating the stability of the foam till 0 mL across the board. To back up the results from the foaming test, the wear test shows that PDMS decreased the wear diameter in all but two circumstances.

7.2.3.1.2 Triethylamine

Triethylamine is an ethylamine oligomer that is used as an anti-foaming agent and anti-friction/anti-wear additive. The triethylamine was tested in concentrations of 250, 500, and 1000 PPM, with the base lubricant being pure water [23]. Water-based lubricants offer cooling properties and low cost, but they also are highly corrosive and have low viscosity. The corrosive properties and low viscosity limit the

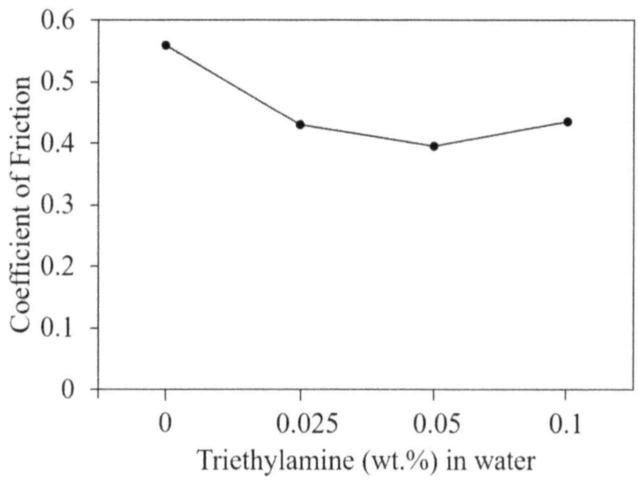

Figure 7.5. Variation in average COF with respect to triethylamine content in water lubricant.

applications that water-based lubricants can be used for. Additives are often used to alter these properties with the purpose of increasing the possible applications. Some of the applications for water-based lubricants are cooling, rolling, and metal-working, where the water can offer both cooling and lubricating properties. The friction coefficient was tested using a tribometer, and the testing material was AISI 52100 steel [23]. The data reported in [23] is plotted and shown in figure 7.5.

The results from the tribotest show that triethylamine worked best at a concentration of 0.05 wt.% and successfully reduced the COF by 30% over that of pure water. The triethylamine also resulted in a significant reduction in the friction coefficient at 0.025 and 0.1 wt.%. For the wear volume test, a mini-track machine was used to mimic the rolling contact and the sliding contact that occurs between the ball and disk. The load was set at 3 N, speed at 20 mm s^{-1}, slide-to-roll ratio or SRR at 10% for a time of 1 h. The volume of wear was then measured using the 'Wyko NT9100 Optical Profiler' [23]. Figure 7.6 displays the results of the wear volume test.

The results of the wear volume test show that the 500 PPM of triethylamine reduced the wear volume by over 59% compared to lubrication with just pure water. The results suggest that triethylamine is a successful additive in terms of anti-foaming, anti-friction, and anti-wear.

7.2.3.1.3 Pribelance™

Pribelance™ is dipropylene glycol isobornyl that offers low toxicity and works well as a co-surfactant similar to long-chain alcohols [24]. In the article, 'Properties of a new hydrotrope hydrophobic molecule and its potential applications,' Pribelance™ is tested as an anti-foaming agent. In the test, the base lubricant used was sodium dodecyl sulphate or SDS and tests were performed with 0–600 PPM of Pribelance™. The foaming of the liquid was performed by injecting nitrogen for 15 s.

Based on the results from the foaming test, we can see that Pribelance™ when at 600 PPM reduces the volume of foam by about 75% compared to the SDS without

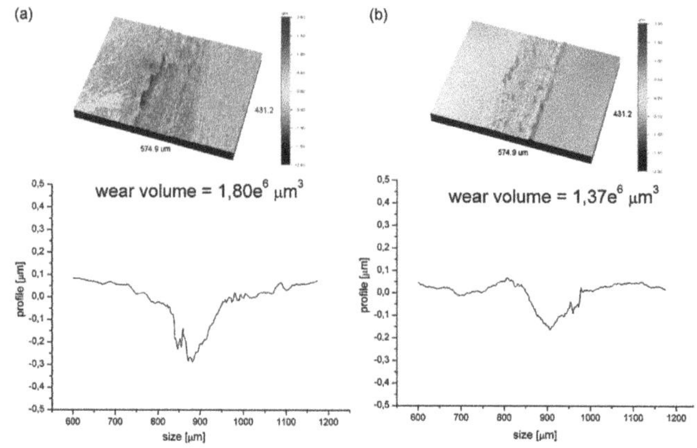

Figure 7.6. Wear volume test results with (a) pure water and (c) 0.05 wt.% triethylamine. Reprinted from [23], copyright (2010), with permission from Elsevier.

any Pribelance™. It can also be noted that there is not a significant reduction in foam volume from 300 to 600 PPM of Pribelance™. Based on these results, we can conclude that Pribelance™ can be used as an effective anti-foaming agent. Pribelance™ has the ability to be dissolved in water when mixed with another, more hydrophilic co-surfactant, which allows it to be a good candidate to be used in cooling lubricants for things like metalworking. Due to its low toxicity, Pribelance™ is also biodegradable, allowing it to be used in processes requiring biodegradable fluids that must be safe for the environment.

7.2.3.2 Anti-oxidants

Anti-oxidants are an essential additive in lubricants in that they are involved in extending the lifetime of the base oils. Oxidation occurs from oxygen coming into contact with the base oil. The rate at which oxidation occurs is known as the onset of oxidation. The onset of oxidation is greatly decreased when contaminants like water and metal are present in the oil and at high temperatures. The onset of oxidation is increased by adding antioxidant additives to the lubricant with the purpose of extending the lubricant's lifetime.

7.2.3.3 Metal deactivators

Metal deactivators share a similar purpose with anti-oxidants in that they are present in lubricants to enhance the life of the lubricant by extending the onset of oxidation. Instead of affecting the rate of oxidation directly, metal deactivators indirectly help to extend the onset of oxidation by reducing the catalytic effect of metals. By reducing the catalytic effect of metals on the oxidation of the lubricant, metal deactivators help to extend the lifetime of the lubricant.

7.2.3.3.1 N,N′-disalicylidene-1,2-diaminopropane (MD)

Dissolved metals often accelerate oxidation in lubricants causing them to decompose. In order to further stabilize the oxidation in the fluid, MD is introduced. The base fluid that MD is tested on in the paper 'Synergistic effect of metal deactivator and antioxidant on oxidation stability of metal contaminated Jatropha biodiesel' is Jatropha methyl ester (JBD), a biodiesel fuel [25]. To begin the testing, a baseline test for the oxidation stability of JBD was taken while contaminants were present. Air was injected into the fuel at 10 L h^{-1} while at 110 °C and the oxidation stability was tested using ASTM D 6751-08 test method EN-14112 [25]. The mean oxidation stability of JBD at 110 °C was 3.95 h [25]. To better stabilize the oxidation of the biodiesel, MD and an antioxidant (AO) were added. MD was added at various concentrations ranging from 5–20 PPM. Throughout the tests of oxidation stability, the concentrations of MD were increased from 5–10–15–20 PPM and during each increase, the oxidation stability continued to increase as well. The best results were found when a 15:1 ratio of AO to MD was used in the biodiesel. Various successfully tested lubricant protective additives are outlined in table 7.5.

7.2.4 Multifunctional additives

Multifunctional additives are additives that serve more than one purpose when added to the base oil. These additives can improve multiple functions of a lubricant such as anti-wear, extreme pressure, anti-rust, viscosity index, antioxidant, anti-friction, and more. The additives discussed in this chapter were the topic of multiple different studies to determine the tribological performance when compared to common lubricants. A list of the additives, testing conditions, and their testing properties is shown in table 7.6. The study of multifunctional additives can be beneficial in decreasing the wear and corrosion of moving parts. They can also be used to reduce the environmental impacts of the additives used today that are not biofriendly.

7.2.4.1 Lipoate esters

Many lubricants that are used today can be harmful to the environment if there is a leak because they are not biodegradable and are usually petroleum-based. Biresaw *et al* [26] investigated the tribological performance of using four lipoate ester additives with polyalphaolefin and sunflower oil base oil. The tribological characteristics that were tested were extreme pressure, anti-wear, viscosity index, and antioxidant properties. A four-ball tester with the ASTM D2783 guidelines was used for extreme pressure testing by increasing the load for 10 s per load.

During the extreme pressure test, Biresaw *et al* [26] found that the weld point of the lipote ester blends increased from 5% to 20% based on which additive was tested. A petroleum-based polysulfide and two enzymatically synthesized lipoate glycerides were compared to the lipoate esters for their extreme pressure performance. The lipoate esters performed very comparably to the lipoate glycerides, with the largest difference being about 200 kgf at 2 w/w% in high oleic sunflower oil (HOSuO).

The ASTM D4172 guidelines were followed while using a four-ball tribometer to study the anti-wear properties of the lipoate esters. The wear scar diameter and the COF of the additives were tested with increasing wt.% of each additive. There was an increase in the COF with the addition of 1% of the additives and a decrease for

Table 7.5. Performance of the lubricant protective additives.

Additives	Base lubricant	Addition content	Testing conditions, material, load, velocity	Tribological properties	Purpose	References
PDMS	HEES, tractor oil, circulating oil, synthetic transmission oil (SAE 75W-90), hydrodynamic transmission oil, neat metalworking oil	15–30 PPM	Foaming test (75, 93.5, 25 °C), Wear test (392 N, 75 °C, 1 h, 1200/1800/2200 RPM	Decreased foaming tendency up to 540 mL, Decrease in wear scar diameter 5%–30% at 2200 RPM	Anti-foaming	[22]
Triethylamine	Pure water	250, 500, 1000 PPM	Tribometer, mini-track machine (MTM), room temp, 20 mm s^{-1}, 3 N, SRR 10%, 1 h	Decreased COF up to 30%, decreased wear volume up to 59%	Anti-foaming	[23]
Pribelance™	Sodium dodecyl sulfate (SDS)	0–600 PPM	Foaming test where nitrogen injected for 15 s	Reduced volume of foam up to 75%	Anti-foaming	[24]
Lipoate Esters	High oleic sunflower oil and PAO	0–20 wt.%	Differential scanning calorimetry: O_2 500 ± 25 psig, 100 ± 10 mL min^{-1}	Increased onset of oxidation temperature	Antioxidant	[26]
ZDDP	Group IV PAO	2 wt.%	Ball-on-disk, steel/steel pure sliding, unidirectional, 1.02 m s^{-1}, max 1.5 GPa, 120 min	When water is involved, will result in higher wear if lubricant contains ZDDP	Antioxidant	[27]
MDA	JP-5 jet and diesel jet fuel	5 PPM	MS and UV absorption detection	Dissolved copper	Metal	
deactivator	[28]					
MD	Jatropha methyl ester	5–20 PPM	Oxidation stability: ASTM D 6751-08 test method EN-14112O_2 injected at 10 L h^{-1} at 110 °C	Helped reduce the amount of antioxidant requirement to meet oxidation stability requirement, best AO:MD ratio of 15:1	Metal	
deactivator	[25]					

Table 7.6. Performance of multifunctional additives in various base oils.

Additives	Base oil	Wt.% /volume	Testing conditions, material, load, velocity	Tribology properties	Purpose	References
Boron (B)-containing thiophosphite derivative (BTP)	Water–glycol hydraulic fluid	<3 wt.%	**Four-ball machine (GB/T 3142-82):** 1450 rpm, 10 s, room temp., 12.7 mm GCr15 steel. *Anti-wear*: 1200 rpm, 30 min, room temp. **Oscillating friction and wear tester:** 20 Hz, 1 mm amplitude, room temp. **Copper Strip Corrosion Test (GB/T 5095-85):** Immersed in liquid for 180 min at 50 °C **Anti-rust Test (GB/T 11143-2008):** Immersed at 60 °C for 24 h.	Base liquid viscosity at 40 °C: 49.54 mm^2/a—pour point: 40 °C	Anti-wear Extreme pressure Anti-rust Friction reducer Anti-corrosion	[33]
Lipoate esters	HOSuO (High oleic sunflower oil) and PAO6 (polyalphaolefin)	0–20 wt.%	**Four-ball machine:** *Anti-wear*: (ASTM D4172-94) *Extreme pressure*: (ASTM D4172-94) **Pressurized differential scanning calorimetry:** Flow rate: 100 mL min^{-1} oxygen to 500 psig	Lipotate esters increased the density of HOSuO at 40 °C and increased the kinematic viscosity at 40 and 100 °C.	Extreme pressure Viscosity index Antioxidant	[26]
Alkylamino dsoamylphos phorodithioates	Neutral grade 150 neutral mineral base oil	0–2 wt.%	**Four-ball machine:** *Anti-wear*: 196 N load, 55 °C, 1800 rpm for 60 min	Load-carrying capacity was improved with higher dosages	Antioxidant Anti-wear	[29]

			Pressure: weld load from 981 N to 2206 N **Pressure differential scanning calorimetry:** *Antioxidant*: temperature from 100 °C to −350 °C, 10 °C min^{-1}, and oxygen flow rate of 60–80 mL min^{-1}. *Anti-friction*: oscillating friction and wear test	Wear reduction of 45%–65% Weld load increased by 40%–80% Secondary amines preparation showed better performance than primary amines.	Extreme pressure Anti-friction	
Difluoroboron derivative	Mineral oil 150 N with 4 wt.% dispersant PIBSI	0–10 wt.%	**Four-ball machine:** GCr15 steel balls with a 60 HRC hardness, 0.032 µm surface roughness, 12.7 mm ball diameter, 490 N *Extreme pressure*: 1450 rpm for 10 s. *Anti-wear*: 75 °C for 60 min, 1200 rpm **Pressure differential scanning calorimetry:**	Metal-free, phosphorous-free	Antioxidant Extreme pressure Anti-wear	[30]

(*Continued*)

Table 7.6. (*Continued*)

Additives	Base oil	Wt.% /volume	Testing conditions, material, load, velocity	Tribology properties	Purpose	References
			Antioxidant: 180 °C and 150 psi oxygen pressure			
Borate ester containing nitrogen with titanium dibutyldithio carbamates	150 SN mineral base oil	0–1 wt.%	**Four-ball machine:** 1450 rpm, GCr15 12.7 mm steel balls, HRC of 59–71. *Anti-wear*: 392, 490, 588 N load for 60 min, *Extreme pressure*: ASTM D2783-88	TiDCC with BNO can have a negative impact on performance. Improved anti-wear synergism with no impact on extreme pressure performance. Reduced phosphorus content.	Anti-wear Load carrying	[31]
Organic borate ester containing nitrogen with Sulfurized titanate	150 SN mineral base oil	0–1 wt.%	**Four-ball machine:** 1450 rpm, GCr15 12.7 mm steel balls, HRC of 59–71. *Anti-wear*: 392, 490, 588 N load for 60 min *Extreme pressure*: ASTM D2783-88	TiC with BNO improved performance. Improved anti-wear synergism with no impact on extreme pressure performance. Reduced phosphorus content	Anti-wear Load-carrying	[31]
Thiadiazole derivatives	Colza oil	0–1.0 wt.%	**Four-ball machine:** GCr15 steel balls at room temp.	Disubstituted derivatives fail at <0.1 wt.%	Load-carrying	[32]

			Friction and wear: 1450 rpm, 392 N load, for 30 min *Extreme pressure*: 1760 rpm for 10 s.	Disubstituted polysulfur derivatives fail at <0.1 wt.%	Weld-load capacity Extreme pressure	
Protic ionic liquid	Poly alpha olefin	0–1.0 wt.%	**Reciprocating sliding test:** GCr15 steel balls at 75 °C, 100 N load, 25 Hz, 30 min **Pressurized differential scanning calorimetry: ASTM D6186-08 (2013), 100 rpm at 30° and bath at 150 °C, 600 kPa**	HDEHP is partially responsible for better tribological performance. NDPA increased the oxidation by being a radical scavenger.	Friction reduction Anti-wear Antioxidant Thermal stability	[34]
Aliphatic acid containing phosphite ionic liquids (SAPN, PAPN, LAPN, OAPN)	Vegetable oil (RSO) Mineral oil (5CST)	0–3 wt.%	**MS-10 J four-ball tester:** 40 kg load at 1450 rpm for 30 min at room temp. **Extreme pressure:** GB12583-98, 1760 rpm for 10 s.	Differences in phosphate species content can be attributed to the friction reduction and anti-wear performance	Low wear volume, anti-wear, friction reduction, extreme pressure	[35]
Carbon quantum dots (CQDs-N)	Bio-based castor oil	0–2 wt.%	**Four-ball tester:** varying load from 196 N to 588 N, 1200 rpm for 60 min at 75 °C **Pressure differential scanning calorimetry:** 50 °C to 170 °C for 2 min until a pressure of 3.5 MPa was reached	The COF was reduced from 0.1 to 0.04 with 0.4 wt.% of CQDs-N with 0.4 wt.% of CQDs-N in the castor oil. The wear scar diameter was also reduced by 40% with 0.4 wt.% of CQDs-N in the castor oil.	Anti-wear Antioxidant Stability Load-carrying	[36]

two of the four additives at 5% addition. The wear scar diameter showed similar results to the COF test with an increase in the wear scar diameter. It increased with 1% additive addition and produced similar wear scar diameter at 0% or slightly smaller values.

The viscosity index of the additives increased slightly at a 1% addition but with a 5% addition, increased by 8–53, with 2-ethylhexyl having the highest increase. Isostearyl stayed the same at 1% addition of the additive and decreased by 14 at 5%.

To test the antioxidant properties, a PDSC was used according to ASTM D6186. The antioxidant properties were evaluated based on the onset temperature and the peak temperature. The onset temperature increased with each higher percentage of the additive, thus showing that the oxidation stability can be improved with the additives. The peak temperature results showed an increase as the additive concentration also increased. It should be noted that isostearyl did not have as significant a peak temperature increase as the rest of the additives, but the cause of this is unknown.

7.2.4.2 *Alkylamino dsoamylphosphorodithioates*

The performance of a common lubricant, metallic dialkyldithiophosphates, called ZDDP was compared to eight ashless alkylamino phosphorodithioates [29]. The alkylamino phosphorodithioates additives are labeled as NPS-1 to NPS-8. The additives were tested for antioxidant performance, anti-wear, extreme pressure, copper corrosion, and anti-friction to determine if the performance improved versus the ZDDP performance characteristics. ZDDP often cannot be used with some types of metals because it produces ash. A four-ball tester was used to test the extreme pressure and anti-wear characteristics of the additives using ASTM D2783. Antioxidant characteristics were tested using differential scanning calorimetry. An oscillating friction and wear test apparatus were used to study the anti-friction properties. The copper corrosion was tested in reference to the ASTM D-130 method. The additives showed wear reduction from 45% to 65% depending on the additive. The weld load showed improvement by 40%–80%. The antioxidant properties improved greatly with higher concentrations. The anti-friction properties were not significantly improved in comparison to ZDDP. The highest COF difference from ZDDP was only about 0.02 less. The research proved that the synthesized alkylamino phosphorodithioate additive performs better than metallic phosphorodiothioates in some tribological characteristics.

7.2.4.3 *Difluoroboron derivative*

A multifunctional lubricant additive called difluoroboron derivative was tested against a common lubricant additive called ZDDP [30]. The multifunctional additive, ZDDP releases sulfated ash that can decrease the life of mechanical systems. For this reason, ZDDP often has strict limitations and restrictions against its use. The difluoroboron derivative was tested for its tribological performance characteristics including extreme pressure, antioxidation, and anti-wear. The extreme pressure and anti-wear capabilities were tested using a four-ball machine per GB/T 3142-82 testing procedures. The antioxidant properties were tested using a pressure differential scanning calorimetry. Using difluoroboron derivatives improved the oxidation induction time by 30%. The nonseizure load of the difluoroboron derivative was improved by approximately 300 newtons from the base oil nonseizure load. The wear scar diameter decreased by 30% when using 5 μmol g^{-1} of the additive. Any further increase or decrease in the additive concentration did not improve the anti-wear capabilities. Difluoroboron derivative decreased the wear volume by 59%, 64%, and 73% when analyzed against ZDDP. The difluoroboron derivative additive improved the tribological performance of the base lubricant and is metal-free and phosphorous-free. This suggests that it could be a good replacement for ZDDP.

7.2.4.4 Borate ester containing nitrogen with titanium dibutyldithiocarbamates

The multifunctional lubricant additive, titanium dibutyldithiocarbamates (TiDDC) with nitrogen borate ester (BNO) was tested for its anti-wear, extreme pressure, and load-carrying capacity in a base mineral oil [31]. ZDDP is one of the most popular multifunctional lubricant additives, but it contains phosphorus which can lead to a shortened lifespan of automotive parts. To test the anti-wear properties, loads of 392, 490, and 588 N were used with TiDDC and BNO separately and together. The additives were tested according to ASTM D4172–82 using a four-ball tester using 12.7 mm, GCr15 bearing steel at a speed of 1450 rpm and 20 °C.

The additive TiDDC showed improved anti-wear capabilities when compared to ZDDP with higher loads. However, TiDCC with BNO showed a decrease in the anti-wear performance. The highest reduction of the wear scar was seen with 1.0%–0.75% TiDCC mixed with 1.0%–0.5% BNO. ZDDP performed the worst out of all of the concentrations and combinations, only being better than the base oil.

The maximum nonseizure load and the weld load are improved with the addition of BNO and TiDCC in comparison to the 150 SN base. The results are seen in figure 7.7 where it is noted that the best load-carrying performance was achieved at a mixture of 1.0% TiDCC and 1.0% BNO. The testing showed that the phosphorus in ZDDP can be eliminated or reduced when it is replaced with boron or TiDCC with BNO. The research suggests that ZDDP is a poor performance additive when

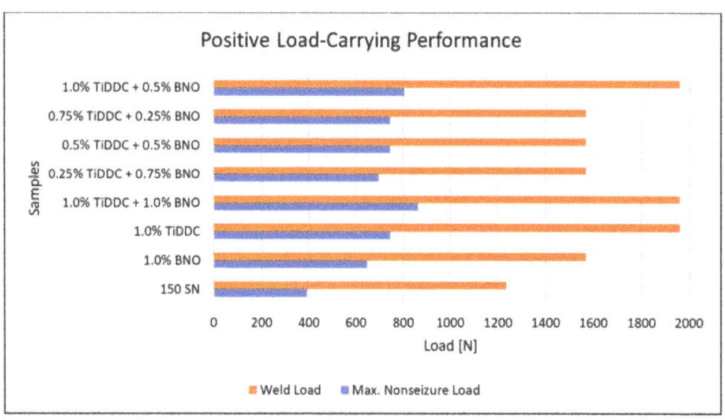

Figure 7.7. Positive load-carrying performance using BNO and TiDCC data extracted from [31]. Reprinted from [31], copyright (2018), with permission from Elsevier.

compared to TiDCC and/or BNO. This conclusion is beneficial to any industry that uses high performance lubricants that want to use more environmentally friendly additives.

7.2.4.5 Borate ester containing nitrogen with sulfurized titanate

An additive called sulfurized titanate (TiS) was combined with borate ester containing nitrogen (BNO) to test for its improved tribological performance [31]. The performance characteristics that were evaluated were the weld load, extreme pressure, and anti-wear. The additive was tested in a 100 °C 150 SN mineral base oil and had a viscosity of 5.1 $mm^2 s^{-1}$.

The anti-wear properties were examined with a four-ball tester with GCr15 bearing steel balls at 1450 rpm and 20 °C. The loads that were used in the experiment were 588, 490, and 392 N. The wear scar diameter was reduced with the addition of BNO and TiS separately as well as combined. By adding TiS and BNO, it can increase the load-carrying capacity compared to just the base oil. TiS improved the maximum nonseizure load more than BNO, but the weld load was improved equally between the two additives. At 1.0% of BNO and TiS, P_B and P_D were improved the most with values of 921 N and 1961 N, respectively. At 0.5% TiS and 1.0% BNO, the weld load was also at its highest value of 1961 N. The research done by Xin Xu *et al* [20] proved that the tribological performance using TiS and BNO was improved with the base oil.

7.2.4.6 Thiadiazole derivatives

Multiple long-chain additives called dimercaptothiadiazole derivatives were tested for their tribological properties, including extreme pressure and anti-wear [32]. The additives were dimercaptothiadiazole based which are monosubstituded derivates termed as MDD (without a branched chain), MTD (with branched chains), and disubstituted derivates termed as DDD (without a branched chain), DTD (with branched chains). These are tested in vegetable oil through the use of a four-ball tester. The tribofilm chemistry and thermal properties were tested using x-ray absorption spectroscopy for tribological performance improvements. With increasing concern for the effect of lubricants on the environment, industry professionals are starting to switch from mineral oils to synthetic lubricants and vegetable oils. The research done to improve the capabilities of environmentally friendly lubricants through the use of additives is very important for a clean energy system.

The tests were conducted using a four-ball tester set at 1450 rpm, 392 loads, and a 30 min testing period. The MTD derivative was better at reducing the friction coefficient in the vegetable oil except at 1.0 wt.% concentration. At 1 wt.% concentration, MTD has a friction coefficient of 0.85 which was slightly higher than that of MDD. It is clear that the DTD with branched chains reduced the friction coefficient more than DDD. The disubstituted derivative with branched chains had the highest reduction in the friction coefficient at a 1.0 wt.% concentration. The friction coefficient was reduced as the concentrations increased for both derivatives except at a concentration of 0.1 wt.%, which increased significantly. The results of the anti-wear testing for the disubstituted polysulfur derivative with branched chains (PDTD) and disubstituted polysulfur derivative without branched chains (PDDD) had varying results based on the wt.% of the additive. PDDD reduced the COF the most with 0.3 wt.% concentration, however, with greater or less concentration, the additive showed a higher COF.

7.3 Conclusions

Overall, surface protective additives play a major role in lubrication stability within oil and contact surfaces. Each additive evaluated played a part in anti-wear/scuff but had specific properties that made them ideal in certain environments. The ILs (trihexyltetradecylphosphonim bis(2-ethylhexyl) phosphate, and polyaniline) possessed excellent thermal stability and decreased the amount of thermal decomposition of the oil, which then resulted in less soot. The absence of soot and debris results in a cleaner oil which reduces the amount of wear and scuff, which some byproducts bring. Detergents and dispersants work in a similar way to remove the insoluble metallic salts, such as zinc-based byproducts when ZDDP is used. ZDDP requires detergents and dispersants due to the high level of zinc thermal decomposition that harms the contact surface. These additives either absorb or disperse the byproducts which then increases the lubrication performance. Surface protection is important in gasoline and diesel engines which require lubrication that can sustain high temperature and constant high-pressure contact. The more development and innovation in surface protecting additives, the higher the performance of these common engines.

Functionalized polymers were found to postpone the inevitable mixture lubrication by absorbing and forming high viscosity surface films. These surface films allow surfaces to not be in full contact in the presence of low entrainment speed conditions. The functionalized polymer has potential in a variety of mechanical systems where friction due to rubbing surfaces negatively affects the overall function of the system. Through the research done analyzing EVA copolymer as an additive to PKME oil, researchers were able to conclude that the inclusion of EVA copolymer had the reaction of significantly increasing the kinematic viscosity of the PKME oil.

Overall, the performance of various lubricant protective additives were evaluated that consisted of anti-foaming agents (PDMS, triethylamine, Pribelance™), 2 anti-oxidants (MoDTP, NPS), and a metal deactivator (MD)). The anti-foaming agents successfully helped to reduce the volume of foam formed while also decreasing the stability of the foam. PDMS decreased the foaming tendency up to 540 mL and decreased the wear scar diameter by 5%–30% at 2200 RPM. Triethylamine decreased the COF by up to 30% and decreased the wear volume by up to 59%. Pribelance™ reduced the overall volume of foam up to 75%. The anti-oxidants helped to increase the oxidation temperature and lengthen the onset of oxidation.

Also, various multifunctional additives were examined and compared to either the base oil or other additives such as ZDDP. The additives showed improvement in performance, surface protection, and lubricant protection. The lipoate ester additives showed an improvement in the antioxidant and viscosity properties. They performed poorly when tested for their anti-wear properties but improved their extreme pressure performance. The alkylamino dsoamylphosphoroditioate additives improved the wear reduction by 45%–65% and weld load by 40%–80% but didn't show anti-friction improvement. When tested against ZDDP, difluoroboron derivative improved the oxidation induction time by 30%, decreased the wear scar diameter by 30%, and improved the nonseizure load by 300 N. TiDDC with BNO as well as TiS with BNO improved the weld and nonseizure load when compared to ZDDP but showed poor anti-wear capabilities. The thiadiazole derivative study compared monosubstituted and disubstituted derivatives and showed that the tribological performance of the colza oil could be improved.

References

[1] Zhang J and Spikes H 2016 On the mechanism of ZDDP antiwear film formation *Tribol. Lett.* **63** 1–15

[2] Qu J, Luo H, Chi M, Ma C, Blau P J and Dai S *et al* 2014 Comparison of an oil-miscible ionic liquid and ZDDP as a lubricant anti-wear additive *Tribol. Int.* **71** 88–97

[3] Cao Z, Xia Y and Chen C 2018 Fabrication of novel ionic liquids-doped polyaniline as lubricant additive for anti-corrosion and tribological properties *Tribol. Int.* **120** 446–54

[4] Cizaire L, Martin J M, Gresser E, Dinh N T and Heau C 2004 Tribochemistry of overbased calcium detergents studied by ToF-SIMS and other surface analyses *Tribol. Lett.* **17** 715–21

[5] Kontou A, Southby M, Morgan N and Spikes H A 2018 Influence of dispersant and ZDDP on soot wear *Tribol. Lett.* **66** 157

[6] Furtado L B, Nascimento R C, Seidl P R, Guimarães M J O C, Costa L M and Rocha J C *et al* 2019 Eco-friendly corrosion inhibitors based on cashew nut shell liquid (CNSL) for acidizing fluids *J. Mol. Liq.* **284** 393–404

[7] Spikes H 2015 Friction modifier additives *Tribol. Lett.* **60** 1–26

[8] Dandan Muhammad A, Samion S, Azman Nurul F, Mohd Zawawi F, Abdul Hamid Mohd K and Musa Mohamad N 2019 Performance of polymeric viscosity improver as bio-lubricant additives *Int. J. Struct. Integr.* **10** 634–43

[9] Soriano N U, Migo V P and Matsumura M 2006 Ozonized vegetable oil as pour point depressant for neat biodiesel *Fuel* **85** 25–31

[10] Asadauskas S and Erhan S Z 1999 Depression of pour points of vegetable oils by blending with diluents used for biodegradable lubricants *J. Am. Oil Chem. Soc.* **76** 313–6

[11] Xue Y, Zhao Z, Xu G, Lian X, Yang C and Zhao W *et al* 2016 Effect of poly-alpha-olefin pour point depressant on cold flow properties of waste cooking oil biodiesel blends *Fuel* **184** 110–7

[12] Faujdar E, Negi H, Singh R K and Varshney V K 2020 Study on biodegradable poly (α-olefins-co-α-pinene) architectures as pour point depressant and viscosity index improver additive for lubricating oils *J. Polym. Environ.* **28** 3019–27

[13] Zhao Z, Xue Y, Xu G, Zhou J, Lian X and Liu P *et al* 2017 Effect of the nano-hybrid pour point depressants on the cold flow properties of diesel fuel *Fuel* **193** 65–71

[14] Reeves C J and Menezes P L 2017 Evaluation of boron nitride particles on the tribological performance of avocado and canola oil for energy conservation and sustainability *Int. J. Adv. Manuf. Technol.* **89** 3475–86

[15] Kasar A K, Reeves C J and Menezes P L 2022 The effect of particulate additive mixtures on the tribological performance of phosphonium-based ionic liquid lubricants *Tribol. Int.* **165** 107300

[16] Alves S M, Barros B S, Trajano M F, Ribeiro K S B and Moura E 2013 Tribological behavior of vegetable oil-based lubricants with nanoparticles of oxides in boundary lubrication conditions *Tribol. Int.* **65** 28–36

[17] You J and Fenfang Y 2010 Tribological performance of two potential environmentally friendly ashless vegetable oil additives *China Petrol. Process. Petrochem. Technol.* **12** 43

[18] Li Z, Xu C, Xiao G, Zhang J, Chen Z and Yi M 2018 Lubrication performance of graphene as lubricant additive in 4-n-pentyl-4′-cyanobiphyl liquid crystal (5CB) for steel/steel contacts *Materials* **11** 2110

[19] Xie H, Jiang B, He J, Xia X and Pan F 2016 Lubrication performance of MoS_2 and SiO_2 nanoparticles as lubricant additives in magnesium alloy-steel contacts *Tribol. Int.* **93** 63–70

[20] Nan F, Xu Y, Xu B, Gao F, Wu Y and Tang X 2014 Effect of natural attapulgite powders as lubrication additive on the friction and wear performance of a steel tribo-pair *Appl. Surf. Sci.* **307** 86–91

[21] Guo M-f, Cai Z-b, Zhang Z-c and Zhu M-h 2015 Characterization and lubrication performance of diesel soot nanoparticles as oil lubricant additives *RSC Adv.* **5** 101965–74

[22] Prolic T C and Lepusic A 2012 Effect of foaming on the antiwear properties of lubricating oils *Goriva Maziva.* **51** 38

[23] Tomala A, Karpinska A, Werner W S M, Olver A and Störi H 2010 Tribological properties of additives for water-based lubricants *Wear* **269** 804–10

[24] Schnell E, Touraud D, Gick R and Kunz W 2008 Properties of a new hydrotrope hydrophobic molecule and its potential applications *Int. J. Cosmet. Sci.* **30** 347–51

[25] Sarin A, Arora R, Singh N P, Sarin R, Malhotra R K and Sharma M *et al* 2010 Synergistic effect of metal deactivator and antioxidant on oxidation stability of metal contaminated Jatropha biodiesel *Energy* **35** 2333–7

[26] Biresaw G, Compton D, Evans K and Bantchev G B 2016 Lipoate ester multifunctional lubricant additives *Ind. Eng. Chem. Res.* **55** 373–83

[27] Cen H, Morina A, Neville A, Pasaribu R and Nedelcu I 2012 Effect of water on ZDDP anti-wear performance and related tribochemistry in lubricated steel/steel pure sliding contacts *Tribol. Int.* **56** 47–57

[28] Loegel T N, Morris R E and Leska I 2017 Detection and quantification of metal deactivator additive in jet and diesel fuel by liquid chromatography *Energy Fuels* **31** 3629–34

[29] Sarin R, Tuli D K, Martin V, Rai M M and Bhatnagar A K 1997 Development of N, P and S-containing multifunctional additives for lubricants *Lubr. Eng.* **53** 21

[30] Wang S, Yu S, Feng J and Liu S 2019 Multifunctional lubricant additive based on difluoroboron derivatives of a diphenylamine antioxidant *RSC Adv.* **9** 35059–67

[31] Xu X, Hu J-Q, Xie F, Guo L, Ma J and Yang S-Z 2018 The synergistic antiwear performances of organic titanium compounds containing sulfur with borate ester additive *J. Spectrosc.* **2018** 2576896

[32] Chen H, Yan J, Ren T, Zhao Y and Zheng L 2012 Tribological behavior of some long-chain dimercaptothiadiazole derivatives as multifunctional lubricant additives in vegetable oil and investigation of their tribochemistry using XANES *Tribol. Lett.* **45** 465–76

[33] Wang J, Wang J, Li C, Zhao G and Wang X 2011 A high-performance multifunctional lubricant additive for water–glycol hydraulic fluid *Tribol. Lett.* **43** 235–45

[34] Jiang C, Wang Y, Su H, Li W, Lou W and Wang X 2019 Synthesis and evaluation of a protic ionic liquid as a multifunctional lubricant additive *Friction* **8** 568–76
[35] Zheng G, Ding T, Huang Y, Zheng L and Ren T 2018 Fatty acid based phosphite ionic liquids as multifunctional lubricant additives in mineral oil and refined vegetable oil *Tribol. Int.* **123** 316–24
[36] Ye M, Cai T, Zhao L, Liu D and Liu S 2019 Covalently attached strategy to modulate surface of carbon quantum dots: towards effectively multifunctional lubricant additives in polar and apolar base fluids *Tribol. Int.* **136** 349–59

Chapter 8

Environmentally friendly cutting fluids

Ashish K Kasar, Alex Kandaras and Pradeep L Menezes

For the development of new engineering components via high-speed cutting, cutting fluids play an important role in machinery. The cutting fluid decreases the temperature at the cutting surfaces by reducing the friction between the tool and workpiece to enhance the life of the tool. However, during the cutting process, the cutting fluid forms mist and micro-droplets, which then cause environment, soil, and groundwater contamination and thus negatively impact human and animal health. The commonly used cutting fluids are mineral (petroleum-based), vegetable and animal oils. This chapter discusses the applicability of environmentally friendly cutting fluids with respect to their chemistry, physical, and chemical properties. The tribological performance of the environmentally friendly cutting fluids is discussed in terms of tool wear and minimum lubrication. Additionally, economic viability is also presented to increase their usage and develop sustainability.

8.1 Introduction

Increased ecological awareness has created imminent demand for a reduction in the use of traditional lubricants (e.g., petroleum-based mineral oils). Mineral oils face the threat of exponential decrease in common availability (petroleum being a non-renewable resource), high toxicity, and poor biodegradation rate. The rate at which a drop of mineral oil in water sees ultimate degradation is close to a full century, and mineral oil that has been improperly disposed off can prove detrimental to aquatic life and inhibit plant growth [1]. Mineral oils produce carbon, nitrogen, and sulfur oxides, causing further concern to be placed upon their negative effects on local air quality and their increasing contribution to global warming on a larger scale [2]. Given the increasing concern for the industrial output of ecologically toxic chemical byproducts, the study of environmentally friendly or 'green' lubricants has become a recent foray into industrial manufacturing in the pursuit of economic and ecological sustainability.

In lieu of unrestrained acquisition of limited, environmentally harmful mineral oils, the main pursuit of such green lubricants has been largely focused on the

doi:10.1088/978-0-7503-3435-8ch8

utilization of processed vegetable oils, liquid or semi-solid plant products comprised of glycerides, and fatty acids occurring organically within the plant structure. In contrast to mineral oils, vegetable oils meet sustainability criteria in their renewability, biodegradability, and nontoxicity to both humans and the world-at-large. Considering vegetable oils as a lubricant base oil provides benefits in the shape of a high flash point, high tool life (due to wetting tendency resulting in less friction), and a high viscosity index, which leaves the door open for development over a wider range of desired tooling temperatures [3]. However, vegetable oils retain lower operating temperature limitations when compared to traditional lubricants and exhibit poor oxidation stability and poor protection against corrosion. As with traditional lubricants, green lubricants can attempt to overcome such shortcomings with additives through typical commercial additives, such as zinc dialkyl dithiophosphate (ZDDP), a common anti-wear additive in motor oils [4].

Lubricants are composed of base oil (e.g., petroleum, synthetic, natural oils) at $\geqslant 85\%$ and performance enhancing additives. The majority of base oils used in the production of industrial lubricants are produced during conventional crude oil processing. This is due largely to the fact that mineral oils are commonly manufactured at scale with a lower price than seen in synthetic or natural oils [5]. Oils obtained from the standard crude oil refining process are referred to as mineral oils. The established and continued use of such mineral oils in industrial lubrication processes, including their proposition as a cutting fluid, has an established negative impact on the environment and health. Over the years, it has become more and more commonplace to try and mitigate maligned ecological effects by producing oil mixtures that improve mineral oil properties by combining them with natural oils [6]. Similarly, eco-friendly refining additives are being developed more frequently, favoring the environment [7].

In this chapter, studies are presented on the lubricants used in the industrial metal machining processes. Lubricants used in the machining of metals (metal cutting fluids) serve to extract heat from the tool and workpiece, reduce friction (thereby reducing wear and increasing tool life and production rate), increase quality of surface finish, and carry chips away from the workpiece [8]. The efficacy of various green metal cutting fluids will be examined upon such traditional parameters.

Further exploration of the demand for a switch to green lubricants is presented herein. The focus of this chapter is to explore the utilization of green lubricants in the realm of industrial machining of metals, metal cutting. The chapter also details a thorough exposition between mineral oil and vegetable oils as a proposed lubricant base oil through an environmental, tribological, and economic lens. Tribological assessment includes assessment of vegetable oils as lubricant bases and the introduction of green additives on performance. A contemporary retrospective will serve to establish a basis on which discussion of the forward mobility towards industrial eco-sustainability can occur.

8.2 Environmental sustainability

Environmental sustainability, in reference to industrial manufacturing, denotes the ability of a given process or resource to maintain ecological integrity. For all purposes, we can assume a lubricant's environmental sustainability implicitly refers to its capacity for biodegradation and its toxicity to animal and plant life. Biodegradation refers to a compound's chemical transformation from a complex (and possibly toxic) state to a more simple and common form consisting of the elements carbon, hydrogen, and oxygen. The process is facilitated via substrate utilizing microorganisms and is naturally occurring across all substances placed into the environment. For a given substance to be adequately biodegradable, it must achieve adequate degradation in reference to some standardized time frame [9]. Toxicity is the degree to which a given compound will adversely affect an organism's health given adequate exposure. Biodegradability and toxicity will be the primary metrics of scrutiny for which conventional (mineral oil-based) lubricants and natural (plant-based) lubricants are assessed with regard to their environmental sustainability.

The Organization for Economic Co-operation and Development (OECD) is an international body which introduces policy for the purpose of creating standardized testing procedures for which the biodegradability can be measured. Table 8.1 establishes a set of important OECD testing methods for evaluating the biodegradation capacity among chemicals included in lubricants [10]. In table 8.1, DOC refers to dissolved oxygen content, BOD refers to biological oxygen demand, and COD refers to chemical oxygen demand. BOD tests seek to measure only the biodegradable portion of organic matter in the cutting fluid, while COD tests measure both the oxygen demand for the biodegradable substance and non-biodegradable oxidizable matter. Biodegradation has two major extents [11]:

- Primary biodegradation: Change in chemical and physical properties of the substance caused by micro-organism activity.
- Ultimate biodegradation: Complete utilization of substance resulting in its conversion into methane (CH_4), carbon dioxide (CO_2), water (H_2O), biomass (microbial cellular constituents), and mineral salts.

Environmental sustainability in the manufacturing process has become a large topic of concern surrounding the effects of industrial petrochemical processing due to the carcinogenic status of petroleum aromatic byproducts, including benzene and toluene [12]. Petroleum, a largely non-renewable resource, can manifest negative impacts on both the environment and petrochemical workers in proximity to the oil

Table 8.1. Standard OECD tests for biodegradability in organic substances.

Test method	Name of the test	Parameters evaluated and the standard adopted
OECD screening tests	DOC die-away test	DOC (OECD 301 A)
	Modified sturm test	CO_2 (OECD 301 B)
	MITI test	BOD (OECD 301 C)
	Closed bottle test	BOD/COD (OECD 301 D)
	MOST	DOC (OECD 301 E)
	Manometric respirometry	BOD/COD (OECD 301 F)

vapors. Petroleum has low solubility rates and can be absorbed into soil banks, permanently damaging soil structures and affecting soil permeability. Along with other malignant effects, oil adheres to the roots of plants, forming mucous membranes and causing damage to the plant matter, in turn destroying microbial ecosystems and thus reducing soil productivity and potential crop yields [13]. It has been observed in northern Mexico that soils based around petrochemical processing plants bear pollution rates 10 times higher than the local standard [14]. Mineral oils, which are consumed at an increasingly alarming rate, provide far less sustainability than vegetable oils, which are naturally occurring, biodegradable, and nontoxic [2].

8.2.1 Composition of petroleum

Petroleum is predominately composed of hydrocarbon molecules (~75% by weight), with sulfur, nitrogen and oxygen occurring in subordinate quantities, with trace metals such as V, Ni, Fe, Na, Ca, Cu, and U. The petroleum-based lubricants can be further categorized based on typical hydrocarbons, as follows [12]:

- normal alkanes (n-alkanes) occur in the highest abundance in petroleum in the homologous series of CH_4, CH_3–CH_3, CH_3–$(CH_2)_n$–CH_3;
- branched alkanes (iso-alkanes) occur in relatively large proportions, especially those with methyl groups;
- cycloalkanes which are cyclic hydrocarbons and majority are cyclopentane and cyclohexane derivatives, and;
- aromatics, inclusive of compounds possessing an aromatic or naphthenic ring.

Hydrocarbons present in crude oil are chemically non-polar, meaning petroleum has a characteristic dissolution in water. Only a small proportion of hydrocarbon constituents present in petroleum will be significantly water soluble.

8.2.2 Mineral oil toxicity

Mineral oils are released into the environment as a form of airborne pollutants created during crude oil refinement. Further, when used as a cutting fluid (especially in open cutting applications), they are further spread in the form of oil mist and micro-droplets, which then penetrate soils and contaminate groundwater [15]. This is of the deepest concern in logging and agricultural applications. Among equal concern is the event of improper disposal of mineral oils resulting in the direct introduction of petroleum into the environment and the possibility of transfer vehicles puncturing. Given the lack of solubility, any petroleum improperly dumped into a potable water source carries with it devastating consequences. For every liter of petroleum a water source is exposed to, roughly a million liters of that body become unsuitable to drink [16].

With petroleum use and petrochemical processing come a wide array of hydrocarbon airborne pollutants. As is observed in the air around petrochemical processing facilities, ambient air tests consistently show significant trace amounts of sulfur dioxide, particulate matter (PM0.5, PM2.5 and PM10), polycyclic aromatic hydrocarbons (PAHs), some metals, as well as volatile organic compounds (VOCs), which include well known carcinogenic substances such as 1,3-butadiene, 1,2-dichloroethane and benzene [17]. The common hydrocarbon pollutants are benezes, toluenen, xylene. The chemical and physical properties of these common pollutants can be found in [18]. These common hydrocarbon pollutants are released in petrochemical processing.

Abundance of such carcinogenic compounds has been shown to have adverse health effects on not only those in direct contact to the processing of crude oils, but also those in the immediate area of these petrochemical processing complexes. A history of exposure to ambient air pollution is associated with several adverse outcomes, including acute lower respiratory infections, chronic obstructive pulmonary disease, asthma, cardiovascular diseases, and lung cancer [19–24]. Petroleum applications invariably introduce aromatic hydrocarbons into their surrounding environment, which are difficult to degrade. After accumulating into soil deposits, they can often enter the human body through the food chain [25]. They have strong mutagenic, carcinogenic and teratogenic effects and threaten human lives and health [26].

8.2.3 Composition of vegetable oils

Vegetable oils, in contrast to mineral oils, are generally homogenous in form and consist of simple fatty acids linked to glycerol with an ester bond. These fatty acids are plant specific and may further vary based on the conditions surrounding the crops growth, mainly climate and nutrients available. Understanding the various fatty acids that can be observed, as well as their composition and properties, will allow for further understanding of how different plant oils can be utilized under different manufacturing pretenses [2]:

- Saturated fatty acids: These fatty acids do not have a carbon–carbon double bond in their backbone. They offer oxidation resistance and exhibit a high pour point.
- Monounsaturated fatty acids: These fatty acids have one carbon–carbon double bond in their backbone structure. They offer relatively a low pour point and reactive sites for chemical modification.
- Polyunsaturated fatty acids: These fatty acids have more than one carbon–carbon double bond in their backbone. They offer a relatively low pour point, a reactive site for chemical modification, but are easily susceptible to oxidation.
- Hydroxy fatty acids: These fatty acids have one hydroxyl group (-OH) in their backbone structure. They offer a relatively low pour point and reactive sites for chemical modification.

Vegetable oils are the product of a renewable resource and are characteristically highly biodegradable, eco-friendly, less toxic, high flash point, low volatility, high viscosity index, wide production possibilities, and economical for waste management [27]. The induction of vegetable oils as a lubricant base oil offers CO_2 equilibrium in that the CO_2 released upon vegetable oil processing is equal to that originally taken up by the plant. Given an increased global awareness (and more stringent regulation) on corporate CO_2 emissions, vegetable oils are an intuitive substitute for mineral oils in the era of ecologically friendly substitutes.

On vegetable oil composition, it is found that the triglyceride structure inherent to plant-based oils provides desirable properties for boundary lubrication due to their long and polar fatty acid chains. These fatty acid chains interact with metallic surfaces and provide a high strength lubricant film, which reduces both friction and wear [28]. Table 8.2 provides fatty acid structures of various vegetable oils.

8.2.4 Biodegradation comparison

In a collaborative exploration of the degradation difference in commercially available mineral oil-based cutting fluids (MO) and plant-based bio-cutting fluid (BCF), postgraduate researchers from the Indian Institute of Technology Guwahati and the Sardar Vallabhbhai National Institute of Technology, India, carried out comparative lab studies on a lathe (Make: HMT, Model: NH26) equipped with a commercial tool holder PTGNR 2525 M22. The experiments are performed for a minimum quantity cutting fluid (MQCF) spray environment with BCF and MO [11]. Tests were performed in a free environment with ample oxygen and water (aerobic aquatic biodegradation). BOD and COD measurements were taken over the course of 5 days, and a percentage degradation ratio was established (BOD_5/COD). Results of such tests resulted in findings that clean BCF produced biodegradation rates of 96.67%, whereas conventional mineral oils had a resultant biodegradability of 18.32% [11].

8.3 Advantages of environmentally friendly cutting fluids

In 2005, the global consumption of lubricants was around 38 million metric tons, with around 85% of said lubricants being mineral oils [39]. The improper disposal of mineral oils (as discussed) leads to surface water and groundwater contamination, soil contamination, and air pollution. This has led to an increased interest in sustainable, alternative solutions to these global manufacturing demands. Apart from their inherent ecological benefits, plant-based oils maintain many rheological properties that maintain their assertion as the logical replacement. Table 8.3 established a brief overview of the advantages and disadvantages of vegetable-based lubricants.

For the listed disadvantages, there are established methods of mitigation dealing with the modification of the base oil, being [29]:

- Reformulation of additives: Many studies on the tribological and wear properties of plant-based lubricants revolve around mixing the base oil with some additives, especially anti-wear and extreme pressure additives.
- Chemical modification: Vegetable oil stability can be improved by modifying the chemical structure of the oil. Such modification includes modification of the carboxyl group (esterification/transesterification), and modifications to the fatty acid chain (selective hydrogenation/formation of C–C and C–O bonds, metathesis, etc); and,
- Genetic modification of crop seeds.

To consider a principle upon which vegetable oils serve as a suitable replacement for mineral oils, we must consider the prior established tribological metrics which constitute the success criteria of cutting fluid. In this, one must consider the

Table 8.2. Fatty acid structures of various vegetable oils, adapted from [28–38].

Fatty acid	Soybean	High oleic soybean	Sunflower	Rapeseed	Palm	Pongammia pinnata	Jatropha curcass	Olive	Castor	Neem	Linseed	Coconut
C 14: 0					1.5	7.9		0.1				
C 16: 0	1.5	6	6	9.8	43	4	12–17	7.3		15	5	9
C 16:1	10.4									22		
C 18: 0	4. 30	3	3	1.6	5	2	6.7	2.7	2–3	19	3	2
C 18: 1	24.4	85	17	18.4	40	62	37–63	61	3–5	49	22	7
C 18: 2	51.6	4	74	16.8	10	12	19–41	4.4	3–5	9.5	17	1
C 18: 3	7.7	2	6	6.5				0.5	80–90		52	
C 20: 0			Traces	9.2	0.5	10						
C 20: 1						4						
C 22: 0	0.5		Traces									
C 22: 1				37.7								
Lauric acid												72

Table 8.3. Advantages and disadvantages of vegetable-based lubricants [29, 30, 33, 40, 41].

Advantages (required qualities of metal working fluids)	Disadvantages
High biodegradability	Low thermal stability
Low pollution of the environment	
Compatibility with additives	Oxidative stability
Low production cost	
Wide production possibilities	High freezing points
Low toxicity	
High flash points	Poor corrosion protection
Low volatility	
High viscosity indices	

rheological properties inherent to vegetable oils and how comparable they are to those seen in their conventional predecessor. In an investigation of the rheology of vegetable oils we will be able to see their potential to initiate and maintain a state of hydrodynamic lubrication in practice.

8.3.1 Rheological properties

The pressure viscosity coefficient (PVC) is a parameter that can be utilized, which is critical in defining a base oil lubricating capacity. The PVC is important because it is reflective of the extent of 'thickening' that occurs under high hydrodynamic loading cases [42]. In a study by Biresaw *et al* [43], the PVC of 10 commodity and new crops was determined based experimentally and based on existing literature and compared against the PVC for two petroleum-based oils (polyalpaolefin (PAO) and hexadecane). It was found that the PVC for the vegetable oils explored was of similar magnitude to that exhibited by the two petroleum-based oils (10–20 GPa^{-1}), with PVC values across the different oils increasing with increasing respective oil viscosity, with decreasing PVC values as temperature increase. While the petroleum-based oils are subject to similar temperature and viscosity effects, the amount to which PVC was reduced in raised-temperature petroleum-based oils (specifically, hexadecane) was negligible in comparison to the reduction seen in the vegetable oils. These results affirm concerns over the thermal instability of vegetable oils.

Lubricity of vegetable oils, dependent highly on chemical composition and moreover, the fatty acid composition of the oil, can vary greatly based on both the plant it is synthesized from as well as an array of circumstantial growth variables surrounding crop growth. Table 8.4 lists some physico-chemical properties of various vegetable oils. It can be seen that the kinematic viscosity decreased at 100 °C compared to at 40 °C. Though reductive, this is consistent with the reduction in PVC as described and brings us to our impasse—pure vegetable oils are not a sufficient replacement for mineral oils. However, vegetable oils have a high chemical proclivity towards modification, especially with extreme pressure (EP) additives. While pure vegetable oils are not (alone) a panacea for a non-sustainable industrial practice.

Table 8.4. Physico-chemical properties of various vegetable-based lubricants [28–30, 32, 34, 44, 45].

Properties	Soybean	High oleic soyabean	Sunflower	Rapeseed	Jojoba	Pongammia pinnata	Jatropha curcass	Neem	Castor
Kinematic viscosity @ 40 °C (cSt)	32.93	41.34	40.05	45.6	24.9	43	47.48	68.03	220.6
Kinematic viscosity @ 100 °C (cSt)	8.08	9.02	8.65	10.07	6.43	8.3	8.04	10.14	19.72
Viscosity index	219		206	216	233	172	208	135	220
Saponofication value (mgKOH g^{-1})	189			180	94.69	179	196.8	166	180
Total acid value (mgKOH g^{-1})	0.61	0.12		1.4	1.1	22	3.2	23	1.4
Iodine value (mg L g^{-1})	144	85.9		104	98	78	97	66	87
Pour point (0 °C)	−9		−12	−12	9	−9	0	9	−27
Flash point (0 °C)	240		252	240			240		250

8.4 Plant-based cutting fluids with extreme pressure additive inclusion

Extreme pressure (EP) additives are conventionally added to lubricants under high pressure scenarios, as in gearboxes, to reduce the wear experienced in the relevant machines. The chemical reactivity of extreme pressure additives plays an important role in the reduction of wear and can serve vegetable-based cutting fluids in such a way as to improve their rheological stability. The introduction of EP additives into vegetable-based cutting fluids leads to efficient friction reduction, lower toxicity and biodegradability. In the machining of metals, when metal-to-metal interactions induce higher temperature and pressure conditions, the lubricant film becomes more stressed, leading to a decreased distance between the metal surfaces. In this case scenario, rubbing and adhesion between the tool and the workpiece becomes likely. EP additives' primary function is to protect the component and add additional protection against such adhesive wear from occurring when the base oil cannot maintain a sufficient film thickness. EP additive concentration of the base oil is of scrutiny, as such additives have a greater tendency towards corrosion in reaction with metallic surfaces.

A study by Kuram *et al* [46] sought to evaluate the performance of various vegetable-based cutting fluids over commercial mineral-based cutting fluids in the turning of aluminum alloy [Al 7075-T6]. Testing consisted of canola and sunflower oils, a blended (canola/sunflower) vegetable-based cutting fluid, and a commercial mineral-based lubricant. A total of 8% and 12% of EP additive (Polartech XP 9018 A) were then added to the cutting fluids. Analysis of cutting, feed, and radial forces determined the reduction in tool wear by 1.70%–38.25% using the vegetable-based cutting fluids over the mineral-based cutting fluids. The lowest flank and nose wear values obtained using the blended cutting fluid containing 12% EP additive were 0.09 mm and 0.10 mm, respectively, whereas values for the commercial mineral cutting fluid were 0.18 mm and 0.15 mm. Scanning electron microscope results found flank and nose wear to be the dominant wear modes.

In a study by Ozcelik *et al* [47], the effectiveness of vegetable-based cutting fluids in application of the turning of AISI 304L was observed. Testing consisted of canola and sunflower oils, using commercial semi-synthetic cutting fluid and a commercial mineral-based lubricant as reference for the comparison of finished surface roughness, tool life, cutting/feed forces. Both canola and sunflower oils were tested at 8% and 12% EP additive quantities. It was found that canola oil with 8% EP weight provided the best performance in terms of surface roughness, feed forces, and tool wear, and that surface quality improved by 35.57% under the use of this mixture over the commercial semi-synthetic cutting fluid and 23.69% over the commercial mineral cutting fluid.

A study concerning the optimization of extreme pressure additive inclusion in vegetable base fluids (sesame, canola, coconut) in the process of turning AISI 1040 steel was performed by Satheesh *et al* [48]. Sulfur-based EP additive (HiTEC343) was used in various weights (5%, 10%, 15%), and was chosen due to its high solubility in water, lower viscosity, and its high lubricating ability. Testing was performed at constant depth of cut (0.5 mm) and three levels for each controlled variable: cutting speed (60, 80, 100 m min^{-1}), feed rate (0.14, 0.17, 0.20 mm/rev).

The optimum parameter setting was defined as the combination of testing parameters that resulted in lower cutting force, cutting tool temperature, tool flank wear, and improved surface finish. Optimum performance was found when using coconut oil at an EP inclusion of 10% by weight at 60 m min^{-1} cutting speed and 0.14 mm/rev feed rate. Desirability function analysis revealed cutting fluid base oil to be the most influential variable, followed by cutting speed, feed and %EP additive.

Vegetable oils, having a high affinity to EP additive inclusion, continue to show increased tribological performance in the presence of EP inclusion. Typical performance increase in consistent with a reduction in cutting and feed forces, and an increase in surface finish and anti-wear properties. While the inclusion of EP alone is enough to induce a performance response within vegetable-based cutting fluids, the magnitude to which the performance can actualize potential performance gains is highly dependent on the base oil, EP additive weight, and mechanical cutting parameters. While optimization can be achieved under set machining operations, it should be clear that results possible will likely not be mass-employable and optimal performance by vegetable-based EP cutting is and will be application specific. Vegetable oil cutting fluids, while maintaining interest in the academic sector, still face this hurdle in the conversation of widespread industrial adoption.

8.4.1 Plant-based nanofluids

Nanofluids are comprised of introducing nano-sized particles into a base fluid (lubricant) [49]. These nano-sized particles possess a better thermal conductivity than the base fluid and thus increase the thermal conductivity of the lubricant upon mixture as a result of their surface area [50]. In addition, the inclusion of such nano-sized particles can serve to increase lubricity in situations wherein the lubricity of the base fluid is not enough. One of such situations is in the case of minimum quantity lubrication (MQL), where the cutting fluid is applied to the workpiece at a rate of no more than 100 mL h^{-1} [51].

The nano-sized particles can be either metallic or non-metallic based on the application [52]. However, in machining (when lubricity is critical), it is typical for the particles used to be of some solid lubricant, such as graphite or molybdenum disulphide (MoS_2). The weak van der Waal's forces inherent in these compounds hold interstitial layers together while allowing the sliding of these layers over each other with ease, reducing sliding friction at the tool/chip interface.

The introduction of such constituents into vegetable oil would require adequate treatment of the oils before disposal in a flood lubrication system. In an MQL application, the fluid is almost fully evaporated throughout the process. Hence no such treatment is required [53].

In two independent studies, MoS_2 (in various amounts) was introduced into coconut, sesame and canola oil in an MQL application while turning AISI 1040 steel was scrutinized [51, 54]. Studies evaluated the oils at different amounts of nanoparticle inclusion (0.25, 0.5, 1 wt.%), cutting speed (40, 60, 100 m min^{-1}), and feed rate (0.14, 0.17, 0.2 mm/rev). Padmini *et al* [54] determined that cutting forces, temperature, tool wear, and surface roughness were reduced by 37%, 21%, 44%, and 39%, respectively, when using coconut oil with 0.5% MoS_2 inclusion compared to dry machining. Rapeti *et al* [51] corroborated the aforementioned results and expanded upon ideal cutting parameters for such usage, with optimal machining performance using this coconut oil nanofluid occurring at a cutting speed and feed rate of 40 m min^{-1} and 0.14 mm min^{-1}, respectively.

A study from Mechiri *et al* [55] investigated thermal conductivity and rheological properties of vegetable oil (ground nut) based hybrid nanofluids of Cu–Zn hybrid nanoparticles, via *in situ* synthesis of copper and zinc particles by mechanical alloying with various component proportions. Nanofluids were prepared at 50:50, 75:25 and 25:75 nanoparticle composition with volumetric concentrations at 0.1%, 0.3%, and 0.5%. Thermal conductivity of the nanofluids was measured with a KD2-Pro thermal property analyzer, and viscosity was measured by measuring the viscous drag of the fluid against the rotation of a spindle using a rheometer. An optimal increase in thermal conductivity was found among the 50:50 particle composition group, resulting in an effective thermal conductivity of roughly twice the thermal conductivity of the base fluid, occurring at 60 °C. Increase in viscosity exhibited a positive logarithmic relationship to volumetric concentration of the nanoparticles, with similar increases occurring across the various levels of particle composition. The maximum increase in viscosity occurred at 0.5% volume fraction, with an increase of approximately 11% (4.7 mPa s).

A study comparing the tribological performance of pure vegetable oil and vegetable oil containing Al_2O_3 nano-inclusions in turning AISI 4130 under MQL application was performed by Gupta *et al* [56]. Al_2O_3 solid particles were dispersed into the vegetable oil at a 0.5% weight, with the objective of recording the performance increase of aluminum oxide addition against the pure vegetable oil, as opposed to mineral oil. Testing was performed at three levels for each controlled variable: cutting speed (50, 80, 100 m min^{-1}), feed rate (0.075, 0.1, 0.125 mm/rev), and depth of cut (0.5, 1, 1.5). In observation, it was found that the characteristic surface roughness (R_a) was characteristically lower for the vegetable oil with aluminum oxide inclusions as opposed to the control group, with the two groups exhibiting similar responses in roughness to the different test parameter combinations. At optimal test levels, it was found that the nanofluid resulted in a 27% improvement in surface roughness when compared to pure oil.

A study concerning the introduction of copper oxide (CuO) nanoparticles into an emulsion of canola oil and distilled water for the purpose of MQL lubrication of surface grinding AISI 1045 steel was conducted by Shabgard *et al* [57]. It was posited of high concentration nanofluids with a high depth of cut where the inclusion of metallic nanoparticles acts to fill the asperity valleys on the metal surface, forming a solid lamellar film between the contacting surfaces and triggering the plastic flow of the tips of the tool and the workpiece, and increasing friction. This is corroborated experimentally by Alves *et al* [58], finding that nano oxide inclusions (CuO and ZnO) slightly raised the friction coefficient in vegetable bases (sunflower and soybean), whereas their presence induced a reduction in friction in mineral- and synthetic-based oils. Moreover, the inclusion of oxides and the implication upon wear effects seem to be dependent upon the base oil. With the induced increase of friction seen among the vegetable oil bases, one can conclude that the introduction of oxide nanoparticles into vegetable oils facilitates third body behavior, inducing poor anti-wear properties.

In conclusion, the oxide inclusions increase friction among vegetable base fluids. Existing research points toward utilizing solid lubricants (e.g., molybdenum disulfide, graphite, etc) as nano inclusions for tribological advancements.

8.5 Economic viability

In consideration of the adoption of vegetable oils as a viable alternative to mineral oils, it would be prescient to discuss the transition from the well-established usage of mineral oils. Adoption of the method must not only prove beneficial in essence but must be fiscally viable. In this vein, one could define 'fiscally viable' as the proof of return that would occur some period beyond adoption. This means having the capacity to recoup any initial transition costs due to additional hardware installation/maintenance/operation. Along these lines, proof that the new method (in operation) is cheaper than the status quo would imply that a fiscal return is just a matter of time and magnitude of operations.

In continuation of discourse on vegetable-based nanofluids, Rapeti *et al* [51] detail a cost analysis based upon then current market estimates (2018) to discern the economic viability of switching a mineral-oil-based flood lubricating system to one using coconut oil at various nanoparticle weights (0.25%, 0.5%) bearing utilization of MQL lubricating techniques in the process of turning AISI 1040 steel. Costs were evaluated assuming a machining time of 6 h per day, 5 working days per week, 52 operative weeks per year, single machine. Fluid costs were based upon available literature [59], with costs of industrial micro-sized MoS_2 particles being quoted out (M/s Luoyang Tongrun Info Technology Co., Ltd, China). Breakdown of said analysis is as follows:

MQL w/ vegetable nanofluids:

- The average rate of exchange from INR to USD for the relevant fiscal year (2018) was applied for conversion (1 INR = 0.015 USD).
- Cost of coconut oil ($0.098 per liter) applied at a rate of 10 mL min^{-1} (MQL) = $912.60 per annum.
- The cost of the micro particles as provided by the aforementioned supplier was quoted at $18.75 kg^{-1}. Thus, particles included at 0.25% (25 mg min^{-1}) = $43.88. Cost of industrial grade MoS_2 micro-sized particles + cost of ball milling (1.5 kW) for 24 h = $43.88 + $240 = $283.88.
- Under MQL operations, disposal costs are negligible.

Flood lubrication w/conventional mineral oils:

- Assuming we have a cutting fluid tank of 120 L and 10% daily loss (12 L) with complete replacement every 4.5 weeks, taking the cost of coolant at $3.38 L^{-1} and the cost of water as $7.73/5000 L, cost of the cutting fluid = $766.68.
- Hazard waste disposal cost is estimated to be $180/ton = $810.

Assuming any costs beyond those that differ in lubricating process (power, cutting tools (study shows that vegetable-based nanofluid utilization decreases tool wear in

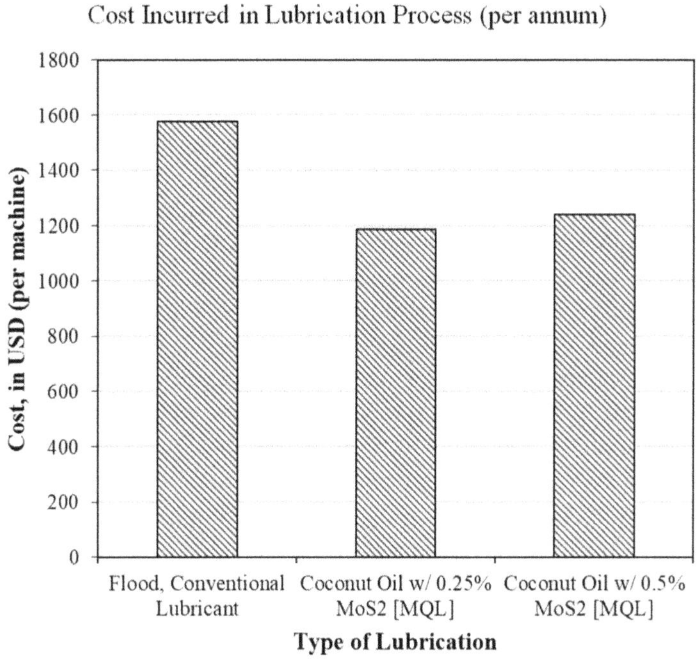

Figure 8.1. Cost incurred in lubrication process (per annum).

comparison to conventional mineral-oil flood lubrication)) are negligible, the cost savings by switching from flood lubrication to MQL = (\$766.68 + \$810) – (\$912.60 + \$283.88) = \$380.20.

As for in the case of 0.5% nanoparticle inclusion (50 mg min^{-1}, 4.68 kg per annum), the cost of buying industrial grade MoS_2 particles and ball milling in-house will be \$327.75. Therefore, yearly savings over conventional flood lubrication will be \$336.33.

Figure 8.1 provides a comparison of flood lubrication and MQL lubrication with vegetable-based cutting fluids at 0.25% and 0.5% MoS_2 inclusion, for a single machine AISI 1040 turning process. Of immediate note is a near 25% reduction in incurred cost when comparing flood lubrication and MQL lubrication methods. With the inclusion of costly nanoparticles, one must also consider that whereas MQL lubrication results in negligible disposal costs, flood lubrication is highly expensive due to the high flow rate and hazardous byproducts treatment costs. Therefore, it can be seen that vegetable-based nanofluids are an economically viable alternative to conventional flood lubrication systems, and it can be intuited that any initial fees incurred in switching from a flood system to one utilizing MQL would be quickly offset in any larger scale operation.

8.6 Conclusions

As concerns over the sustainability of petroleum-based products (including mineral oil-based lubricants and further cutting fluids) continue to grow, so will the demand for a more wholly sustainable replacement. Mineral oils have shown to be detrimental to the environment and workers who have direct exposure to petroleum products and petrochemical processes. Residents in towns housing petrochemical processing plants are shown to experience a higher risk for respiratory diseases, endocrine diseases, skin conditions, and adverse reproductive effects. Petroleum is non-renewable, and poses ecological detriments in that it is non-biodegradable and improper disposal can be retained in soils, water and wildlife, where it inevitably enters a food cycle for which humans can still be subject to trace amounts of petroleum and petroleum byproducts, including notably carcinogenic aromatics (e.g. benzene, toluene). Vegetable oils are, by contrast, readily abundant and renewable by nature. Toxicity to humans in the vegetable oils being explored as replacements are nominal, if not negligible.

Vegetable oils as a base fluid offer an intuitive solution to the proposal of sustainable lubricants. They maintain low ecological threat (renewable, low toxicity), high flash points, and low volatility, among other traits which suit them as a premier candidate. They do pose their own set of challenges for mass application, which include traits such as low thermal stability, high freeze point, poor corrosion protection, etc. However, their ready compatibility with additives makes it a question of what additive sets (and in what proportion) will optimize a vegetable-based lubricant for a given application. Experiments showing the use of extreme pressure additives and nanoparticle inclusions in the turning of aluminum and steel turning, respectively, show that additive modified vegetable-based cutting fluids can reduce tool-workpiece friction, tool wear, cutting forces, and temperature, etc. In an analysis of the use of such vegetable-based cutting fluids, there is an apparent corroboration of the validity of an attempted shift to these environmentally friendly lubricant alternatives.

In the case of single machine turning operations, it was found that utilization of MQL lubricating and coconut oil decreased operating costs considerably over conventional mineral oils lubricants in conjunction with flood lubricating methods. When considering the possible scaling implications of this elementary single unit cost analysis, the case for the economic feasibility of the adaptation of vegetable-based cutting fluids in conjunction with MQL application becomes apparent.

Vegetable-based cutting fluids are an appropriate replacement for conventional mineral oils in that they have the capacity (in proper formulation) to outperform mineral oils on typical tribological metrics of success. Moreover, the possibility of substantiating the transition from mineral oils to vegetable oils in fiscal terms is key to mass adaptability, and the extrapolation of an elementary case may provide optimism in this fight. The promotion of manufacturing processes that are both ecologically sustainable and fiscally viable are key in adaption for the further promotion of an environmentally friendly industrial sector, and the exploration of green alternatives, including environmentally friendly cutting fluids, is a stepping stone in a much longer path to actualizing such ideals.

Acknowledgements

The authors acknowledge the support of the National Science Foundation (Grant No. CMMI-2010205).

References

[1] Boyde S 2002 Green lubricants. Environmental benefits and impacts of lubrication *Green Chem.* **4** 293–307
[2] Menezes P L *et al* 2013 *Tribology for Scientists and Engineers* (Berlin: Springer)
[3] Pettersson A 2007 High-performance base fluids for environmentally adapted lubricants *Tribol. Int.* **40** 638–45
[4] Habereder T, Moore D and Lang M 2009 Eco requirements for lubricant additives *Lubricant Additives* ed M Rudnick (Boca Raton, FL: CRC Press) p 647
[5] Syahir A *et al* 2017 A review on bio-based lubricants and their applications *J. Clean. Prod.* **168** 997–1016
[6] Farfan-Cabrera L I *et al* 2019 Tribological performance of an engine mineral oil blended with a vegetable oil under approached long-term use conditions *SAE Technical Paper* 2019-01-0012
[7] Vasile C *et al* 2017 Comparative analysis of the composition and active property evaluation of certain essential oils to assess their potential applications in active food packaging *Materials* **10** 45
[8] Katna R *et al* 2017 Green manufacturing-performance of a biodegradable cutting fluid *Mater. Manuf. Processes.* **32** 1522–7
[9] Aluyor E O and Ori-Jesu M 2009 Biodegradation of mineral oils—a review *Afr. J. Biotechnol.* **8** 915–20
[10] OECD guidelines for the testing of chemicals, section 2: Effects on biotic systems https://oecd-ilibrary.org/environment/oecd-guidelines-for-the-testing-of-chemicals-section-2-effects-on-biotic-systems_20745761?page=3
[11] Gajrani K K, Ram D and Ravi Sankar M 2017 Biodegradation and hard machining performance comparison of eco-friendly cutting fluid and mineral oil using flood cooling and minimum quantity cutting fluid techniques *J. Clean. Prod.* **165** 1420–35
[12] Posthuma J 1977 The composition of petroleum *RapP.-v. Reun. Cons. Int. Explor. Mer.* **171** 7–16
[13] Tejada M, Hernandez M and Garcia C 2009 Soil restoration using composted plant residues: Effects on soil properties *Soil Tillage Res.* **102** 109–17
[14] Iturbe R *et al* 2008 TPH and PAH concentrations in the subsoil of polyduct segments, oil pipeline pumping stations, and right-of-way pipelines from Central Mexico *Environ. Geol.* **55** 1785–95
[15] Iturbe R *et al* 2005 Subsoil TPH and other petroleum fractions-contamination levels in an oil storage and distribution station in north-central Mexico *Chemosphere.* **61** 1618–31
[16] Nowak P, Kucharska K and Kamiński M 2019 Ecological and health effects of lubricant oils emitted into the environment *Int. J. Environ. Res. Public Health.* **16** 3002
[17] Domingo J L *et al* 2020 Health risks for the population living near petrochemical industrial complexes. 1. Cancer risks: a review of the scientific literature *Environ. Res.* **186** 109495
[18] Wang S *et al* 2021 Advances in research on petroleum biodegradability in soil *Environ. Sci.: Process. Impacts.* **23** 9–27
[19] Sack C *et al* 2017 Air pollution and subclinical interstitial lung disease: the Multi-Ethnic Study of Atherosclerosis (MESA) air–lung study *Eur. Respir. J.* **50** 1700559
[20] Shahadin M S *et al* 2018 Challenges and future direction of molecular research in air pollution-related lung cancers *Lung Cancer* **118** 69–75

[21] Wang C-W *et al* 2019 Association between urinary thiodiglycolic acid level and hepatic function or fibrosis index in school-aged children living near a petrochemical complex *Environ. Pollut.* **244** 648–56
[22] Wang C-W *et al* 2019 Urinary thiodiglycolic acid is associated with increased risk of non-alcoholic fatty liver disease in children living near a petrochemical complex *Environ. Int.* **131** 104978
[23] Bergmann S *et al* 2020 Effect modification of the short-term effects of air pollution on morbidity by season: a systematic review and meta-analysis *Sci. Total Environ.* **716** 136985
[24] Chen X *et al* 2020 Susceptibility of individuals with chronic obstructive pulmonary disease to air pollution exposure in Beijing, China: a case-control panel study (COPDB) *Sci. Total Environ.* **717** 137285
[25] Gildeeva I 1999 Environmental protection during exploration and exploitation of oil and gas fields *Environ. Geosci.* **6** 153–4
[26] Baud-Grasset F, Baud-Grasset S and Safferman S I 1993 Evaluation of the bioremediation of a contaminated soil with phytotoxicity tests *Chemosphere* **26** 1365–74
[27] Mannekote J K *et al* 2018 Environmentally friendly functional fluids from renewable and sustainable sources-a review *Renew. Sustain. Energy Rev.* **81** 1787–801
[28] Siniawski M T *et al* 2007 Influence of fatty acid composition on the tribological performance of two vegetable-based lubricants *J. Synth. Lubr.* **24** 101–10
[29] Shashidhara Y M and Jayaram S R 2010 Vegetable oils as a potential cutting fluid—an evolution *Tribol. Int.* **43** 1073–81
[30] Adhvaryu A and Erhan S 2002 Epoxidized soybean oil as a potential source of high-temperature lubricants *Ind. Crops Prod.* **15** 247–54
[31] Asadauskas S and Erhan S Z 2001 Thin-film test to investigate liquid oxypolymerization of nonvolatile analytes: assessment of vegetable oils and biodegradable lubricants *J. Am. Oil Chem. Soc.* **78** 1029–35
[32] Joseph P, Saxena D and Sharma D 2007 Study of some non-edible vegetable oils of Indian origin for lubricant application *J. Synth. Lubr.* **24** 181–97
[33] Pop L *et al* 2008 Basestock oils for lubricants from mixtures of corn oil and synthetic diesters *J. Am. Oil Chem. Soc.* **85** 71–6
[34] Singh A and Gupta A 2006 Metalworking fluids from vegetable oils *J. Synth. Lubr.* **23** 167–76
[35] Anand O, Mehta J and Prasada T R 1998 Lubricant components from vegetable oils of Indian origin *J. Synth. Lubr.* **15** 97–106
[36] Choi U *et al* 1997 Tribological behavior of some antiwear additives in vegetable oils *Tribol. Int.* **30** 677–83
[37] Jayadas N and Nair K P 2007 Elucidation of the corrosion mechanism of vegetable-oil-based lubricants *J. Tribol.* **129** 419–23
[38] Jayadas N, Nair K and Ajithkumar G 2007 Tribological evaluation of coconut oil as an environment-friendly lubricant *Tribol. Int.* **40** 350–54
[39] Kline and Company I 2006 *Competitive Intelligence for the Global Lubricants Industry, 2004–2014.* (Little Falls, NJ: Kline & Company, Inc.)
[40] Mercurio P, Burns K A and Negri A 2004 Testing the ecotoxicology of vegetable versus mineral based lubricating oils: 1. Degradation rates using tropical marine microbes *Environ. Pollut.* **129** 165–73
[41] Cermak S C and Isbell T A 2003 Synthesis and physical properties of estolide-based functional fluids *Ind. Crops Prod.* **18** 183–96

[42] Sharma B K and Stipanovic A J 2002 Pressure viscosity coefficient of lubricant base oils as estimated by nuclear magnetic resonance spectroscopy *Ind. Eng. Chem. Res.* **41** 4889–98
[43] Biresaw G and Bantchev G B 2013 Pressure viscosity coefficient of vegetable oils *Tribol. Lett.* **49** 501–12
[44] Sivasankaran G *et al* 1988 Jojoba-oil-based two-stroke gasoline engine lubricant *Tribol. Int.* **21** 327–33
[45] John J, Bhattacharya M and Raynor P C 2004 Emulsions containing vegetable oils for cutting fluid application *Colloids Surf. A.* **237** 141–50
[46] Kuram E *et al* 2013 Effects of blended vegetable-based cutting fluids with extreme pressure on tool wear and force components in turning of Al 7075-T6 *Lubr. Sci.* **25** 39–52
[47] Ozcelik B *et al* 2011 Experimental investigations of vegetable based cutting fluids with extreme pressure during turning of AISI 304L *Tribol. Int.* **44** 1864–71
[48] Satheesh Kumar B *et al* 2017 Optimisation of turning AISI 1040 steel with extreme pressure additive in vegetable oil based cutting fluids *Int. J. Automot. Mech. Eng.* **14** 4285–97
[49] Bianco V *et al* 2015 *Heat Transfer Enhancement with Nanofluids* (Boca Raton, FL: CRC Press)
[50] Das S K *et al* 2007 *Nanofluids: Science and Technology* (New York: Wiley)
[51] Rapeti P *et al* 2018 Performance evaluation of vegetable oil based nano cutting fluids in machining using grey relational analysis-a step towards sustainable manufacturing *J. Clean. Prod.* **172** 2862–75
[52] Srikant R *et al* 2014 Nanofluids as a potential solution for minimum quantity lubrication: a review *Proc. Inst. Mech. Eng., Part B: J. Eng. Manuf.* **228** 3–20
[53] Bystrzejewska-Piotrowska G, Golimowski J and Urban P L 2009 Nanoparticles: their potential toxicity, waste and environmental management *Waste Manage. (Oxford)* **29** 2587–95
[54] Padmini R, Krishna P V and Rao G K M 2016 Effectiveness of vegetable oil based nanofluids as potential cutting fluids in turning AISI 1040 steel *Tribol. Int.* **94** 490–501
[55] Mechiri S, Vasu V and Venu A 2017 Gopal, Investigation of thermal conductivity and rheological properties of vegetable oil based hybrid nanofluids containing Cu–Zn hybrid nanoparticles *Exp. Heat Transfer* **30** 205–17
[56] Gupta A *et al* 2020 Comparative performance of pure vegetable oil and Al_2O_3 based vegetable oil during MQL turning of AISI 4130 *Mater. Today Proc.* **28** 1662–6
[57] Shabgard M, Seyedzavvar M and Mohammadpourfard M 2017 Experimental investigation into lubrication properties and mechanism of vegetable-based CuO nanofluid in MQL grinding *Int. J. Adv. Manuf. Technol.* **92** 3807–23
[58] Alves S M *et al* 2013 Tribological behavior of vegetable oil-based lubricants with nanoparticles of oxides in boundary lubrication conditions *Tribol. Int.* **65** 28–36
[59] Amrita M, Srikant R and Sitaramaraju A 2013 Evaluation of cutting fluid with nanoinclusions *J. Nanotechnol. Eng. Med.* **4** 031007

www.ingramcontent.com/pod-product-compliance
Ingram Content Group UK Ltd.
Pitfield, Milton Keynes, MK11 3LW, UK
UKHW051854100926
8556IPUK00005B/618

9 780750 334334